Saba Bebawi is Senior Lecturer in Journalism at the University of Technology Sydney. She holds a PhD from the University of Melbourne, and is the author of *Investigative Journalism in the Arab World* and co-editor of *Social Media and the Politics of Reportage*.

'A much needed approach for the understanding of journalistic discourse in the emerging complexity of globalised news spheres.'
— Ingrid Volkmer, author of *The Global Public Sphere: Public Communication in the Age of Reflective Interdependence*

MEDIA POWER AND GLOBAL TELEVISION NEWS

The Role of Al Jazeera English

Saba Bebawi

BLOOMSBURY ACADEMIC
LONDON • NEW YORK • OXFORD • NEW DELHI • SYDNEY

BLOOMSBURY ACADEMIC
Bloomsbury Publishing Plc
50 Bedford Square, London, WC1B 3DP, UK
1385 Broadway, New York, NY 10018, USA
29 Earlsfort Terrace, Dublin 2, Ireland

BLOOMSBURY, BLOOMSBURY ACADEMIC and the Diana logo
are trademarks of Bloomsbury Publishing Plc

First published in Great Britain by I.B. Tauris 2016
Paperback edition published by Bloomsbury Academic 2021

ISBN: HB: 978-1-7845-3086-0
PB: 978-1-3502-4233-3
ePDF: 978-0-8577-2731-2
eBook: 978-0-8577-2935-4

Typeset in Garamond Three by OKS Prepress Services, Chennai, India

To find out more about our authors and books visit
www.bloomsbury.com and sign up for our newsletters.

For all the journalists who have used their media power to develop and support a democratic 'global public sphere'.

CONTENTS

LIST OF ILLUSTRATIONS

LIST OF ABBREVIATIONS

AFP: Agence France-Presse
ARABSAT: Arab Satellite Communications Organization
AP: Associated Press
BBC: British Broadcasting Corporation
CCTV-9:   China Central Television 9
CNN: Cable News Network
CNNI: Cable News Network International
CDA: Critical Discourse Analysis
DA: Discourse Analysis
IPS: Inter Press Service
ITU: International Telecommunications Union
NWICO: New World Information and Communication Order
Telesur: Televisora del Sur (Television of the South)
UN: United Nations
UNESCO: United Nations Educational, Scientific and Cultural
                Organisation
WSIS: World Summit on the Information Society

ACKNOWLEDGEMENTS

I would like to thank the following people for their support and assistance throughout this project. First, I would like to express my sincere appreciation to Dr David Nolan and Associate Professor Ingrid Volkmer for their insightful advice and input. I would also like to acknowledge the Swinburne Institute for Social Research for assisting in funding this project. I would like to thank my work colleagues at Swinburne University of Technology, Monash University, and Zayed University in Abu Dhabi, throughout the period of this research. Particularly, my deepest gratitude goes to Associate Professor John Arnold for being my mentor and my friend. You have been there every step of my academic career, and this book has been no exception – for this I am eternally grateful.

At a personal level, I would like to thank my friends for offering every form of encouragement throughout the writing of this book. I am greatly thankful to my parents Dr Ali El-Ghul and Fathieh Tijani, and brother Mohammad El-Ghul, to whom I am everlastingly indebted for their relentless support and perceptive feedback. My greatest inspiration has been my daughter Sia who came into my life at the start of this project, and whose sunny persona has been the light directing this study. And finally, my warmest appreciation goes to my life partner, Sam Bebawi, for going out of his way to make this book a priority – you have been a pillar in this journey.

INTRODUCTION

AL JAZEERA ENGLISH AND MEDIA POWER

'The media are the most powerful institutions on earth – more powerful than any bomb, more powerful than any missile'
Listening Post, *Al Jazeera English Broadcast, 25 February 2007*

The above quote broadcast on Al Jazeera English (AJE), as part of a preview of its media watch programme *Listening Post*, reflects a self-conscious awareness of the crucial role that the media play in conflict reporting by likening the media to weapons of war. While this claim is highly dramatised, and perhaps rather hyperbolic, it reflects the power of media institutions and their attempt to control media discourse. This symbolic power, to use John B. Thompson's (1995) term, constructs discourses of which we interpret the relevance of conflicts in various perspectives. In this sense, 'media power' centres on the particular capacity of media to construct social reality, thus providing the symbolic resources through which conflicts are perceived (Couldry, 2000; Boyd-Barrett, 2002). Furthermore, as has been pointed out in various studies, media diplomacy is dependent on the media, where decision makers rely on the news as a source of information. Indeed, Nick Couldry and James Curran argue that 'media power' can be defined as 'a label for the net result of organizing a society's resources so that the media sector has significant independent bargaining power over and against other key sectors (big business, political elites, cultural elites, and so on)' (Couldry and Curran, 2003: 3). This suggests that media power, and its potential contestation, are vital issues for contemporary research.

Despite media power's capacity to affect daily news events, it is arguably even more significant in influencing global crises. The mediation of global conflicts not only impacts how they are conceived by a global audience, but also shapes and determines the dynamics of the conflict itself. On this, Simon Cottle has argued that '[h]ow global crisis reporting serves to reproduce the voices of the powerful or variously gives vent to the range of views and contention that surround, shape and inform crises clearly constitutes an important, though often overlooked, dimension of the media's role in global crisis reporting' (Cottle, 2009: 38). This suggests, he argues, that we need to pay much closer attention to the different ways in which contemporary forms of news reporting 'provide different opportunities for the public elaboration, deliberation and visualization of crises' (Cottle, 2009: 38). This has been particularly evident in the role media, and in particular social media, have recently played in the protests in the Middle East.

The Middle East has been a particular focus of global crisis reporting. Yet, international coverage of these conflicts has historically been presented through a 'Western' perspective. The absence of Arab voices in the 'global public sphere' has created a discursive gap between the Middle East and the rest of the world. AJE might, thus, be regarded as an attempt to bridge this gap by broadcasting discourses from and about the Arab world. Hence the aim of this book is to specifically understand AJE's position in the global news environment for the purpose of identifying the extent to which it addresses this gap between the Arab and global spheres. This book therefore conducts a discursive framing analysis of selected news reports by AJE before and after the Arab Spring protests.

Moreover, this book looks into the proliferation of satellite news over the last two decades which has stimulated substantial debate over the degree to which the field of transnational media serves to extend hegemonic perspectives and power, or enable an increasing number of voices and perspectives to gain representation. Specifically, it seeks to critically consider claims made by AJE, in relation to other global news networks, that suggest it serves to present 'alternative' perspectives on international news. This book argues that AJE is a 'hybrid' news network which is both contesting and supporting discourses presented by dominant global broadcasters such as CNNI and the BBC.

The Rise of Al Jazeera

In general, 'transnational media' refers to cross-border media transmission which allow radio and television channels to broadcast outside the national borders, and 'are usually grouped under the single "transnational" category, which is useful only when distinguishing them from those that remain within national boundaries, as it tends to mask their extraordinary diversity' (Chalaby, 2005: 2). Transnational media, Jean K. Chalaby suggests, can be divided into four categories according to their area of reach, and these are the local, national, world-regional and global reach (Chalaby, 2005: 4). Major international satellite television news networks, such as BBC World and CNNI, may be seen to have a privileged capacity to influence both global news agendas and, in turn, the discursive terms through which the events that such agendas are engaged with. Various scholars, particularly those who have centred on the political economy of global media, have argued that global reporting has been primarily dominated by a handful of news networks and agencies without much representation from other parts of the world (Murdock and Golding, 2005; McChesney, 2003). However, while such critics have suggested that this domination has mainly been from Western news providers, the field of international television news has witnessed the entry of mainly English-language non-Western global broadcasters seeking to both counter-balance and compete with Western news channels. Such examples of satellite television broadcasters include the 'pan-Latin American TV channel Televisora del Sur ("Television of the South", Telesur)' which broadcasts from Venezuela, the

'English-language global television channel, Russia Today (RTTV)', the 'English language network of China Central Television' (CCTV-9) (Thussu, 2007: 12), and AJE which began broadcasting in November 2006 from Qatar.

Al Jazeera Arabic (AJA) began broadcasting to Arabic-speaking audiences in the Middle East. It was founded by the Emir of Qatar, and began broadcasting in 1996 from Doha, Qatar. Media scholarship that has focused on AJA has tended to present it as a democratising influence in the mediated Arab public sphere, as a consequence of its provision of discourses that were not previously openly discussed on any Arab media platform. AJA has launched a number of specialised television channels in addition to its English service, such as the Al Jazeera Sports Channel, Al Jazeera Sports Plus 1 and Al Jazeera Sports Plus 2. Al Jazeera also broadcasts an Arab channel called Al Jazeera Live which airs continuous live political, economic, or cultural events such as news conferences, discussions or meetings. It has also set up a children's channel with the cooperation of the Qatar Foundation for Education, Science and Community Development. In addition, Al Jazeera broadcasts a documentary channel in Arabic and runs two online websites: one in Arabic, aljazeera.net; and one in English, english.aljazeera.net. AJA cooperates with other networks such as Telesur, the Latin American television station, with which it has established an agreement to share training and exchange content (Zayani and Sahraoui, 2007: 164–165). Through such ventures, Al Jazeera has sought to significantly expand its operations beyond the Middle East, and become an increasingly influential, and profitable, global player. Politically, AJA has played a key role as an increasingly prominent voice within the Arab world, and has been positioned as exercising a significant 'democratising' power by challenging previously dominant discourses both within the Arab public sphere and, more recently, the 'global public sphere' (Volkmer, 2007; McNair, 2006).

The inception of AJE came as a result of AJA's strategy to broadcast to a broader English-speaking audience. AJE began broadcasting from Doha in November 2006, operating from 'centres in Kuala Lumpur, London and Washington, as well as 20 other bureaus' (*The Age*, 17 November 2006). Furthermore, as will be discussed in this book, the establishment of AJE was presented by the channel itself as an opportunity to provide an Arab discursive angle on issues and events in the Middle East to a global audience, in addition to counter-balancing the coverage of dominant media. This has been made possible as a result of AJE broadcasting in English, which makes its broadcasts far more accessible to audiences outside the Arab-speaking world. The reason behind this, according to AJE, is to provide the global English-speaking audience with a better understanding of Arab conflicts by broadcasting Arab discourses that are not usually tackled by dominant global news media.

Bridging the Gap

The Middle East has been a particular focus of global crisis reporting. The ongoing Palestinian/Israeli crisis, the war in Iraq and Afghanistan, and recently the rise of the Arab Spring, have all been consistent features in daily world news. Yet, international coverage of these conflicts has historically been presented through a 'Western' perspective. The absence of Arab voices in the 'main' 'global public sphere' has created a discursive gap between the Middle East and the rest of the world. AJE might, thus, be regarded as an attempt to bridge this gap by broadcasting discourses from and about the Arab world. Hence the aim of this book is to understand AJE's position in the global news environment for the purpose of identifying the extent to which it addresses this gap between the Arab and global spheres. The bridging of this gap is imperative in order to supply global audiences with the knowledge needed in building a more comprehensive and multi-perspectival understanding of world events, and in turn forming political opinions as a result of this knowledge, specifically in relation to conflicts happening in the Arab world. Accordingly, this book considers the extent to which AJE is challenging the power of dominant global news networks and/or presenting differing and insightful perspectives in order to provide an Arab discursive angle on political crises in the Middle East that might not match the discourses broadcast by dominant global media.

This study explores how AJE reports on Arab regional conflicts as global news events, whereby I provide an analytic understanding of the nature of its news reporting from its initial stages of operation until post-Arab Spring protests, which mark a significant political shift in the region. The study institutionally situates AJE within the mediated 'global public sphere' through an analysis of the discourses it projects in comparison to those of its global competitors in English-language global satellite news broadcasters, namely CNNI and BBC World. Primarily, this book is concerned with asking whether AJE, as a recent entrant to the field of international news, appears to provide additional and/or alternative mediated discourses on local regional conflicts to a global audience in order to better understand the potential role AJE is playing within the mediated 'global public sphere'. It questions the discursive nature of AJE's news reporting and the role it is playing as a relatively recent entrant to the mediated 'global public sphere'. I do this by comparing AJE's news reporting to CNNI and the BBC as dominant news organisations, while also using these two news networks as case studies for uncovering the nature of their discursive output as dominant news providers within the field of international news and in relation to AJE.

This study, therefore, explores both theoretically and empirically the claims made by AJE itself as providing an 'alternative' voice within the mediated 'global public sphere', and performs a framing analysis of its news reporting in comparison to BBC World and CNNI in order to understand the extent to which it could be labelled as an 'alternative' platform. A comparative approach, between these three transnational news channels as case studies, has been chosen since it is important to understand how specific news stories are covered by each of the three channels; however, this is done through what I describe as a 'contextual' focus on AJE's performance. The reason behind selecting CNNI and BBC World as comparative case studies with AJE is due to the three news networks being 'tough competitors', where the 'BBC and CNN already have Arabic online services [and] [e]ach is keen to carve a niche in the others' market' (Barkho, 2007: 13). For this reason – along with the BBC and CNNI being news networks that AJE sought to emulate as will be discussed later – these two dominant media channels become suitable case studies for a comparative analysis. This book, therefore, seeks to uncover the nature of the global satellite news space and relations of media power through the case study of AJE, in order to understand the extent to which it plays a role in bridging the gap between the Arab sphere and the 'global public sphere'.

CNNI has a long tradition as a 'globalised news channel'. CNNI employs correspondents around the world and, in addition, CNNI exchanges pictures and video feeds from digital cameras and even mobile phones from various unconventional and non-traditional sources, which the channel labels as 'citizen journalism' (Masie, 2006: 74). CNNI also has reciprocal arrangements with other news channels. The content produced in the CNNI newsroom is presented in various formats such as video feeds, streamed video and scrolling headlines at the bottom of the screen (Masie, 2006: 75). The significance of CNNI's news reporting initially began in its ability to break news and provide instant developments on the news story from primary sources. The 'CNN-effect' is a term that 'came to be used to refer to the impact of news media (both domestic and global) in general and not just that of CNN' (Robinson, 2011: 3). Although the concept of the 'CNN-effect' has been debated extensively in the literature, it has been strongly tied to understanding 'the various influences upon the foreign policy decision-making process' (Robinson, 2011: 4). Such political influence would suggest that CNNI's reporting does implicitly advocate a political position. Yet although CNNI is a US-based news network, the channel does not regard itself as carrying an American perspective through its international service (Sullivan, 2002). This, however, will be further explored through the analytic study conducted in this book.

BBC World is also positioned with a focus on 'first-hand coverage of the latest breaking news with a commitment to depth, context and intelligent analysis' (Boaden, 2004). BBC News has 41 bureaux located overseas (Boaden, 2004) and has, over time, experienced an increase in global reach especially with its online presence, whilst its national audience has been in decline (Flood et al., 2011: 221).

Although BBC World is a global broadcaster in the sense that it has a global reach, 'the BBC does operate and act within a British or Western culture and context, and its content will always in some ways reflect this' (Mytton, 2008: 571). Accordingly, it is reasonable to hypothesise that the BBC's news reporting can be expected to advocate a British perspective, particularly when it comes to political crisis reporting. This would also suggest a paradox since, as Flood et al. argue, the BBC 'prescribes impartiality, fairness and balance in news reporting', yet at the same time there is a requirement as a result of it being the British broadcaster 'to serve specified civic values deemed fundamental to British society' (2011: 221). They provide an example of how the BBC's coverage, despite its own rhetoric of impartiality, implicitly advocates particular perspectives when reporting on particular topics such as 'Islam-related reporting'. What this indicates is that the BBC's reporting seems contextual, based on the topics at hand – a point that will be addressed in the final chapter in light of the findings of this study.

In addition to uncovering AJE's discursive practice within the 'global public sphere', the question of whether established conceptualisations of international news flows within the literature, such as the media imperialism thesis, remain valid in the face of the arrival of new players into the field of global satellite news will also be discussed. Accordingly, this book addresses the questions: Does AJE, as an Arab broadcaster in the international television news space, offer distinct and/or challenging discourses that reflect a regional Arab perspective? Accordingly, to what extent can AJE be regarded as a news network that challenges dominant media power through its discursive news output? How can AJE's discursive role be understood within the context of the mediated 'global public sphere'? These questions will be addressed and discussed in detail.

Plan of the Book

In order to address the above research problems, this book is theoretically grounded in an engagement with two areas of academic debate. The first area of debate is centered on the discussion of whether news media developments, with a particular focus on the field of satellite news broadcasting, are enabling the realisation of a 'global public sphere'. An engagement with this first area of debate, which is undertaken in the opening chapter of this book, enables particular positions that have arisen around AJE to be placed within a wider context, through the consideration of both theoretical approaches to the 'public sphere', and different theoretical perspectives on recent developments in international news. This chapter addresses the global satellite news space by discussing debates on the 'global public sphere' and tying concepts drawn from these debates, such as Nancy Fraser's (2007) concept of 'interlocutors', to AJE's possible role in the transnational public sphere. Drawing on Fraser's work, I consider the possibility of it acting as a 'subaltern counterpublic'. This chapter discusses three varying positions on the question of a mediated 'global public sphere': the media imperialism thesis; the expansion of the 'global public sphere'; and what I label as *domination and resistance*. Using Fraser's concept of 'interlocutors', this chapter illustrates these positions with examples of global television news networks in differing national contexts. This discussion aims to provide a clearer understanding of the nature of the field which AJE has entered, and accordingly provide a stronger basis for specifically discussing AJE in relation to media power.

The second chapter looks more specifically into Al Jazeera's development as a news organisation through the *national*, *transnational* and *global* models of the Arab public sphere. It outlines the debates surrounding AJE's launch, as a relatively recent broadcaster in the global news space, through existing scholarly debates on Al Jazeera's historical role in the Arab and 'global' spheres and through AJE's own articulation of its operation. It has been claimed that the significance of AJE lies in its capacity to provide alternative perspectives on issues that are not available elsewhere. It was in recognition of this that this enquiry sought to consider to what degree such claims might be justifiable. Prior to its launch, AJE promoted itself as a channel that offers an 'alternative' source of news, a profile that has been created by AJA.

Al Jazeera has, generally therefore, been known for providing news discourses from an Arab or Muslim perspective that aim to challenge what it sees as dominant discourses coming out of global news networks. This chapter provides a discussion on AJE's self-promotion through newspaper articles, online commentary, staff interviews, and its own broadcasts. It argues that there is a 'tension' in AJE's promotion as a channel that is attempting to balance itself between the Arab and 'global' spheres.

The second area of theoretical debates this book looks into are those surrounding media power. Chapter 3 explores the notion of media power and media power contestation as articulated through academic debate, highlighting that media power is constructed and played out through mediated discursive texts. I refer to media power in this book as 'discursive media power', and it is in this chapter that I address this through the methodological approach used in this study, which is that of Critical Discourse Analysis (CDA). CDA provides a methodological tool for understanding relationships of power embedded in media texts, and has, therefore, been identified as a suitable method of study for assessing the extent to which AJE challenges perspectives of its global rivals. This is based on Fraser's (2007) argument, discussed in the opening chapter, who suggests the need to address not only 'who' is to emerge as participants in the 'global public sphere', but more importantly 'how', and it is only through a discursive analysis that we can understand *how* media power is actually played out.

Chapters 4 and 5 offer an analysis of different selected news stories with the aim of uncovering the degree to which AJE can be considered a news network that could provide counter-discourses before the Arab Spring and post-Arab Spring, and in turn act as a counter-power to dominant broadcasters operating within the 'global public sphere'. Each analytic chapter looks into two news events that were selected for inclusion in this book out of a broader observational study of various news stories during the study period. While a case study approach cannot hope to deliver a representative sample of coverage, a detailed focus on the forms of discourse analysed provides suggestive grounds for considering the degree to which AJE presents an 'alternative' perspective to its major international rivals. Significantly, the news stories analysed focus on events in the Middle East. Chapter 3 tackles two stories which coincided with the start of AJE's broadcasts in 2007. The first news story is the kidnapping of the BBC Gaza correspondent, Alan Johnston, who was taken by a Palestinian group in return for the release of Muslim prisoners from foreign jails on 12 March 2007, and released on 4 July 2007. This chapter also includes the story of the six Bulgarian medical workers who were sentenced to death in Libya for allegedly infecting over 400 Libyan children with the AIDS virus in February 1999, and who were released in July 2007. This chapter captures examples of AJE's discursive news reporting within its first year of operation. The findings of the analysis indicate that AJE's news output can be considered a 'hybrid' practice that neither strongly enforces nor contests dominant media power.

Chapter 4 looks into two news events that occurred in relation to the Arab Spring uprisings in both Egypt and Syria. The first of these news events is the sentencing of 528 members of the Muslim Brotherhood, on 24 March 2014 in Egypt's criminal court in Minya, for participating in an attack on a police station in mid-August 2013 in which a member of the police force was killed. The second news event is that of the UN Security Council's veto on 19 July 2012 in New York, which failed to adopt a UN resolution on Syria due to the Russian Federation and China vetoing a British-sponsored resolution that would have imposed economic sanctions against the Syrian government for failing to carry out a peace plan. Overall this chapter offers an insight into how AJE's textual output could be understood eight years after its launch. This chapter argues that during this time, its practice of hybrid reporting has remained unchanged.

The final chapter of this book sheds light on the broader research questions this book is concerned with addressing, and on the relevant positions of scholarly debate pertaining to those discussed in the book. This final chapter serves to provide a more complex picture than can be adequately addressed through binary frameworks of mainstream/alternative, publics/counter-publics, and flows/contra-flows discussed in the first part of the book. Accordingly, here I argue that AJE's discursive reporting could be better articulated through a hybrid of mainstream and alternative journalistic practice. In turn, what this hybrid practice suggests is that the inclusion of additional voices from previously unrepresented parts of the world into the mediated 'global public sphere', while significant, does not guarantee that 'contestation' will be a subsequent result. Moreover, it was also notable from the analysis that the reporting of BBC World and CNNI, as dominant global broadcasters, is not itself uniform in its mediated representation. Accordingly, this chapter concludes by questioning the established binaries found in media literature. Accordingly, here I argue that AJE's discursive reporting could be better articulated through a consideration a contextual practice where AJE is reporting according to various journalistic factors which influence its output such as, among others, *routines of news sourcing*, the influence of the *culture of journalism* on reporting practices, and the nature of the *media organisation* itself and its relationship with its audiences. Accordingly, these elements which I label in this book as *forces of discursive media power*, contribute to the construction of particular forms of discursive output, and possibly contend, in the 'global public sphere' of satellite news.

Contribution and Approach

The contribution this book makes to existing literature is significant in many ways. The first is that it provides a critical account of AJE's performance by highlighting the complexities and binaries existing in, what I label as, AJE's hybrid reporting. This study, therefore, offers a critical assessment of AJE's news output, which is not fully supportive of other published studies that describe AJE as 'a voice to the voiceless' (Painter, 2008; El-Nawawy and Powers, 2010). Although, this book acknowledges instances of this, it however argues that this might not always be the case, and that AJE practices a more hybrid form of discursive reporting.

Another significant contribution this book makes is that it also argues the problematic binaries continuously stated in global media literature such as North/South, mainstream/alternative, flow/contra-flow. This book shows how these binaries discursively overlap, and in the case of AJE, media institutions cannot be described as either one or the other. In addition, this study illustrates how dominant news organisations, in this case BBC World and CNNI, also differ in their media output thus emphasising the limitations presented when labelling them as 'Western' media.

This book studies AJE through a discursive analysis of its news output due to the gap in the literature on understanding *how* AJE's news reporting can be understood and, in turn, conceptualised. Naomi Sakr argues that in order to test a station's claim to alternative practice, it needs to demonstrate that it 'consciously pursued a distinctive news agenda' (Sakr, 2007a: 106). Therefore, as I argue in this book, the categorisation of whether a media channel is 'alternative' or not needs to be conducted through the network's discursive content rather than how its management, editors and staff claim it to be. There are many studies that have looked into AJE's institutional make up, interviewing journalists and editors, in addition to assessing the impact of AJE and how it is perceived by its viewers (see El-Nawawy and Powers, 2010). Other studies, such as Ustad Figenschou (2010), have conducted a quantitative content analysis of AJE's news. These studies have been useful, however the need for this book stemmed from the necessity to understand AJE's actual reporting through a qualitative approach, in order to test whether its output matches its institutional claims. The importance, therefore, of this study lies in its focus on the detailed discursive analysis of AJE's reporting. In this book, I argue that it is not only 'who' participates in the mediated 'global public sphere' that matters, but more importantly 'how'; and it is only through a discursive analysis that we can understand *how* media power is actually played out.

Another significance of this book lies in the fact that this research has followed AJE from the commencement of its broadcasts until recently, which made it possible for an observational analysis to be conducted from the station's initial broadcasts. It provides an analysis of AJE's work from the start of its launch, in addition to recent events post-Arab Spring protests, thus providing an understanding on how

AJE's reporting has developed and/or changed. It is worth noting that this study paves the way for other studies by emphasising AJE's hybrid news reporting, thus inviting further research on uncovering the reasons behind this in order to understand how editorial and reporting decisions were made and why.

Overall, this book offers a critical study of AJE's discursive performance that does not necessarily support some previous studies of the network that are rather celebratory in tone. Through the study of AJE, this work makes a contribution to considering the relative strengths of different positions and perspectives in light of a consideration of empirical evidence, enabling a grounded reflection on the significance of developments within the transforming field of international satellite news.

CHAPTER 1

NEWS NETWORKS AND THE 'GLOBAL PUBLIC SPHERE'

Over the last 30 years, since the launch of CNN in 1980, satellite television channels broadcasting 24-hour news have mushroomed around the world and news networks have come to define the global news agenda around us. Focusing on humanitarian crises, political tensions, and breaking news, these international networks have sought to bring to the forefront issues and discourses that make up the mediated 'global public sphere'. This opening chapter seeks to shed some light on the question of whether or to what extent developments in satellite news broadcasting – such as the inclusion of the Arab satellite channel Al Jazeera English (AJE) – have contributed to the realisation, or enlargement, of a mediated 'global public sphere' (Cottle, 2009). In order to do so, however, it is first necessary to critically engage with different theoretical definitions of, and approaches to such a mediated 'public sphere', in terms of the different ways in which the historical reality is understood as, on one hand, an 'actually existing' historical reality, and on the other as a conceptual basis for considering the role of media in facilitating democratic communication. Following this conceptual overview, the chapter goes on to consider the relation between these conceptions and contemporary debates on satellite news, which often deploy the concept of the public sphere, in order to understand the extent to which a mediated 'global public sphere' has been realised. Through a critical review of these debates, the chapter seeks to provide a mapping of the various debates and positions in relation to an 'actually existing contemporary public sphere'. In doing so, it also seeks to develop connections with chapters to follow, by considering a framework for addressing alternative media content in terms of the degree to which it 'contests media power'.

The Public Sphere – Perspectives and Definitions

In the last decades of the twentieth century, criticisms of the global domination of 'Western' media and the power they hold became entrenched in debates surrounding global news flows. However, with the development of satellite technology many developing countries have launched their own media operations in their own languages, targeting both local audience segments and diasporic audiences globally, leading to an increased focus on the significance of 'diasporic public spheres' (Thussu, 2005). More recently, the rise of English-language transnational satellite news channels targeting a global audience has meant that media from various parts of the world can, at least theoretically, now cater for audiences that are culturally different from their own. Such a phenomenon leads to the consideration of potentially new possibilities of the expansion of a 'global public sphere', as a result of the recent entrants into transnational broadcasting from parts of the world that were not previously represented (Demers, 2002; Lull, 2007; Volkmer, 1999), such as AJE.

Jürgen Habermas has offered an historic account of the public sphere which has been a basis for the emergence of numerous re-conceptualisations, hence it is necessary to provide a brief account of Habermas's model of the public sphere as a background to understanding the evolution of a possible 'global public sphere'. Habermas conceptualised the power of deliberation within the public sphere through the role of national media in eighteenth-century Europe. Habermas's public sphere offers an important historical and theoretical model where he presents an eighteenth century ideal of a public space, informed by principles of equality and openness, and oriented towards the development of rationally debated, consensual positions on issues of collective concern. This ideal was generated by the 'bourgeois public' who, as a reading public that claimed new spaces and publications for exchanging ideas, opinions and arguments, operated independently from state authorities as a sphere of 'civil society'. Habermas's account of the public sphere is outlined in his 1962 book, *The Structural Transformation of the Public Sphere*, where he discusses the emergence of a liberal bourgeois public in the eighteenth century as comprising the educated and influential elite of society in Western Europe (particularly England, France and Germany). Drawing on Marxist class analysis, Habermas labelled this section of society as the 'bourgeois' stratum, comprised of 'the merchants, bankers, entrepreneurs, and manufacturers [...] This stratum of "bourgeois" was the real carrier of the public, which from the outset was a reading public' (Habermas, 1992: 23). Habermas describes this public sphere as follows:

> The bourgeois public sphere may be conceived above all as the sphere of private people come together as a public; they soon claimed the public sphere regulated from above against the public authorities themselves, to engage them in a debate over the general rules governing relations in the basically privatized but publicly relevant sphere of commodity exchange and social labour (Habermas, 1992: 27).

Habermas describes the public sphere as a 'specific domain', representative of 'the public' in contrast to both the state and the private sphere, that emerged in opposition to the authorities of the eighteenth century. According to Habermas, its emergence marked a separation of the private and public domains and the rise of the 'public sphere of civil society' (Habermas, 1992: 23). This historic process of the rise of the public sphere was facilitated by the emergence of early finance and trade capitalism where 'the elements of a new social order were taking shape' (Habermas, 1992: 14). Merchants were no longer confined by the town and instead were associated with enterprising companies, thus creating a 'bourgeois' stratum of society whose interests were distinct from those of state rulers, despite the fact they consolidated states through the payment of taxation (Habermas, 1992: 23–24). Habermas describes how this 'bourgeois public' met in salons and coffee houses which constituted them as a public sphere, where there were shared certain criteria that, he suggests, were largely shared across otherwise the distinct social contexts of Western Europe (Habermas, 1992: 36–37). He enumerates these criteria as comprising the following common elements. Firstly, there was a bracketing of social status within these spaces of public debate, and an assumption of equality amongst all its members. Secondly, in these coffee houses and salons topics of debate were addressed that had not previously been publicly discussed – with particular reference to issues concerning the conduct of state and church authorities – that had themselves largely delimited topics and parameters of discussion prior to the establishment of the bourgeois public sphere. Thirdly, aware of the fact that they belonged to a larger public, bourgeois participants in this newly formed sphere of civil debate also sought to circulate their debates and discourses through publications, in journals such as the Addison and Steele's *Spectator* which 'was the leading journal for circulating the cultural principles of liberal-bourgeois politics' (McGuigan, 1996: 25).

Habermas argues that the public sphere has been in decline, relating this to the commercialisation of mass media which transformed the public sphere once more to an arena where the public no longer has access to contribute to public debate, but rather become consumers of 'representative publicity' (Habermas, 1992: 12–13). The news, therefore, followed the development of the business enterprise by similarly adopting characteristics of 'concentration and centralization' (Habermas, 1992: 186). Habermas also relates the commercialisation of news, arguing that '[t]echnological development in the means of transmission of news (after the telegraph and the telephone came the wireless telegraph and telephone and shortwave and radio) [which] has in part hastened and in part made possible the organizational unification and economic interlocking of the press' (Habermas, 1992: 186–7). The power that the mass media hold in shaping public opinion, Habermas claimed, led to the media taking on the role formerly exercised by the feudal sovereign. Habermas describes this as the process of 'refeudalization', as follows:

> In the measure that is shaped by public relations, the public sphere of civil society again takes on feudal features. The 'suppliers' display a showy pomp before customers ready to follow. Publicity imitates the kind of aura proper to the personal prestige and supernatural authority once bestowed by the kind of publicity involved in representation (Habermas, 1992: 195).

A key aspect in Habermas's argument, therefore, is that media power 'emerged [as] a new sort of influence' which was 'used for purposes of manipulation' (Habermas, 1992a: 437), where he describes this as 'the manipulative deployment of media power to procure mass loyalty, consumer demand, and "compliance" with systematic imperatives' (Habermas, 1992a: 452). Such manipulation, according to Habermas, is employed through the use of discourses which 'imply power structures that are not only hidden but systematically latent, that is, structurally concealed from their participants' (Habermas, 1992a: 478).

This historic account of the public sphere has, however, provoked various criticisms (for example Thompson, 1995; Calhoun, 1992; Eley, 1992). One of the key arguments is that Habermas's description of the public sphere tends to underplay, if not entirely dismiss, other civil society movements active within the general public that were also critical of the authorities, yet did not see themselves as part of the 'bourgeois public' (Thompson, 1995). As Jim McGuigan argues, the 'radical, alternative public sphere, represented by popular newspapers such as *The Poor Man's Guardian,* had a powerful impact upon the bourgeois public sphere, not only in opposition but also as a source of progressive ideas' (McGuigan, 1996: 26). A second point of criticism raised by Thompson is that Habermas fails to mention other periodicals that were representative of the time, thus tending to exclude the existence of other debates that were circulated. The third issue which Thompson identifies relates to the limitations of the bourgeois public sphere itself, which was dominated by wealthy, educated and male citizens, thus excluding other potential members of the public. This criticism of Habermas's public sphere mainly relates to class- and gender-based exclusions that ensured that, while the public sphere may have 'in principle' been inclusive, in practice it was a space of debate that was, in part, defined as much by its exclusions as the principles of inclusivity highlighted by Habermas.

These observations serve, in addition, to highlight another point of criticism raised about Habermas's model: its presumption that 'the public sphere' might exist as a singular space of rational debate animated by a desire to achieve consensus via the 'force of the better argument' (Habermas, 1990: 158–159). If, on one hand, as others claim (Negt and Kluge, 1993; Landes, 1988) multiple public spheres have existed, and if on the other criteria for judging arguments were, at least in part, determined by the particular class and gender characteristics of the bourgeois public sphere, then it follows that both the descriptive accuracy and the normative force of Habermas's account of the public sphere come into question. This has led others to develop models of the public sphere which seek to encompass a greater level of pluralism.

If, as discussed above, questions can be raised about the degree to which the eighteenth century bourgeois public sphere could be defined as a shared space of rational argumentation, it can be claimed that the disparity between the Habermasian ideal and contemporary reality has only widened. The validity of Habermas's model as a basis for engaging with today's media environment has, for example, been strongly questioned by Todd Gitlin who argues that 'the public sphere' is labelled as '*the* sphere, not *a* sphere' (Gitlin, 1998: 168), and thus ignores the existence of smaller active and interactive spheres. Here 'the unitary public sphere', Gitlin argues, 'is weak, riddled with anxiety and self-doubt, but distinct communities of information and participation are multiplying, robust and brimming with self confidence' (Gitlin, 1998: 170). To support his case, Gitlin offers examples from the media industry which demonstrate the existence of a multiplicity of publics, particularly cable television which offers 'targeted channels for targeted audiences' (Gitlin, 1998: 171), thus reflecting the existence of demographically distinct audiences whose different interests and tastes are targeted by media producers. In addition to questioning the empirical accuracy of Habermas's model, Gitlin also stresses that pluralism within a multicultural society provides an image of a 'segmented unity' (Gitlin, 1998: 173), such that the theoretical validity of a singular 'public sphere' of debate may also appear questionable on democratic grounds.

While Gitlin does acknowledge the emergence of an increasingly global 'interconnected world', and draws on the example of 'citizen groups which are organized by political affinity' (Gitlin, 1998: 171), he remains concerned about whether the trend towards interactive 'sphericules' could pave the way for the materialisation of a space of shared democratic dialogue. As he states, '[w]hat is not clear is that the proliferation and lubrication of publics contributes to the creation of a public – an active democratic encounter of citizens who reach across their social and ideological differences to establish a common agenda of concern and to debate rival approaches' (Gitlin, 1998: 173). Similarly, through his interest in forms of communicative public engagement and marketing used by various political groups that have developed sophisticated strategies of public communication, Clifford Bob (2008) describes the global sphere of civil society as a fragmented and politically fractious one, arguing (rather pessimistically) that rather than being oriented towards consensus, the contemporary public sphere is a conflictual space characterised by the attempts of various groups, seeking to gain attention and influence. He argues that the common belief that 'new voices add to the marketplace of ideas' is commonly aligned with assumptions that global civil society consists of harmonious and 'like-minded' groups, when in fact quite the reverse is true (Bob, 2008: 201). In this way, Bob follows Gitlin in suggesting that, in contrast to Habermas's model, the fragmentation of public spheres appears to offer limited prospects for the achievement of shared spaces of consensually oriented debate. He also suggests that, although it might produce plural domains of discourse, 'civil society's conflictive nature does not in itself make it democratic', since '[g]lobal civil society remains primarily an arena of elite, not mass, politics', in large part due to increased access to communicative and financial resources available to elite interest groups (Bob, 2008: 201).

While Bob thus presents an image of contemporary public life as one where fractiousness and fragmentation belies idealistic conceptions of shared public debate, others have drawn on Gitlin's concept of 'sphericules' to consider spaces of debate that exist alongside what persists as the dominant public sphere of mass media representation in particular national contexts. Stuart Cunningham, for example, uses Gitlin's concept to consider distinctive spaces of communication of ethnic minorities that exist alongside, and contribute to, 'the' public sphere of 'host' societies (Cunningham, 2004: 153). Cunningham draws on this model to draw attention to 'ethno-specific global mediatized communities' which can be termed 'sphericules' since they:

> share many of the characteristics of the classically conceived public sphere – they provide a central site for public communication in globally dispersed communities, stage communal difference and discord productively, and work to articulate insider ethno-specific identities – which are by definition 'multi-national', even global – to the wider 'host' environments (Cunningham, 2004: 153).

However, quite contrary to Gitlin and Bob who are pessimistic when it comes to the consideration of a dominant public sphere, others highlight public communication across communities of difference. Cunningham, for example, refers to diasporic communities relative to national cultures as being part of 'wider "host" environments'. In this respect, where Gitlin and Bob appear sceptical about the possibility of shared spaces of public representation, Cunningham presents 'sphericules' as contributors to a 'unity in diversity'. In all of these cases, however, the concept of 'sphericules' departs from Habermas's model because it acknowledges the existence of proliferating spaces of identity with their own distinctive characteristics, norms and communicative forms. This is important both as a basis of alternative description (the mediated public sphere cannot be conceptualised as a singular space), and in terms of its theoretical implications (one cannot appeal to universal norms of rationality as these are not, and are unlikely to be, universally shared across such a plurality of publics). In this respect, the recognition of 'sphericules' is a significant departure from Habermas's model of the public sphere, since it seeks to theorise how other voices serve to constitute a multiplicity of public spheres that may, or may not, produce mutual dialogue and debate.

By contrast to such models of fragmentation and 'public sphericules', a third model has sought to engage with how the existence of a shared dominant public sphere co-exists with, and may be transformed by, 'counterpublic spheres'. Such a model has been elaborated in the work of Nancy Fraser (1993), who also proposes this model, in contrast to Habermas's framework, as offering a more empirically valid and politically productive means of theorising the public sphere. Fraser draws particularly on arguments that demonstrate that Habermas's bourgeois public was never the 'main' or only public, since 'competing counterpublics' were in operation well before the late nineteenth and twentieth centuries, including 'nationalist publics, popular peasant publics, elite women's publics, and working-class publics' (Fraser, 1993: 7). In addition, Fraser argues that the existence of a multiplicity of publics, rather than the ideal of a single, unitary public sphere as articulated by Habermas, offers a more democratic aspiration, since social inequalities typically do not allow subordinate groups within a single public sphere to have a strong voice as a result of different levels of power (Fraser, 1993: 14). Particularly, and more recently, Fraser (2007a) critiques Habermas's conceptualisation of the public sphere through national boundaries, especially in relation to his focus on national media, which neglects increasingly emerging global media. In regards to this, Fraser argues:

> *Structural transformation* associated the public sphere with modern media that, in enabling communication across distance, could knit spatially dispersed interlocutors into a public. Tacitly, however, Habermas territorialized publicity by focusing on national media, especially the national press and

national broadcasting. Thus, he implicitly assumed a national communications infrastructure, contained by a Westphalian state (Fraser, 2007a: 10).

Here Fraser argues towards a more transnational conceptualisation of the public sphere, a point I address further below.

Nonetheless, unlike Gitlin who questions the existence of a unitary public sphere, Fraser also refers to a 'dominant public sphere'. However, in contrast to both Gitlin's fragmented account of 'sphericules' and Cunningham's account of sphericules operating within a 'host' public environment, Fraser's model provides a theoretical image of smaller spheres operating 'outside' the dominant public sphere. Fraser proposes the label of 'subaltern counterpublics' to describe publics which represent 'parallel discursive arenas where members of subordinated social groups invent and circulate counter-discourses, so as to formulate oppositional interpretations of their identities, interests, and needs' (Fraser, 1993: 14). The term subaltern, she explains, refers to 'subordinated social groups' such as 'women, workers, peoples of color, and gays and lesbians' who have 'found it advantageous to constitute alternative publics' (Fraser, 1993: 14). The term counterpublics, on the other hand, are publics which contest and compete with other publics, thus contributing to a representative sphere rather than a 'single, comprehensive, overarching public sphere' (Fraser, 1993: 14).

Fraser perceives subaltern counterpublics as formed on the basis of two principles. The first is that these spaces act as a means of identity where they are based on the interests and concerns of the members of that particular public. Such spaces operate through a set of shared discourses, and enable engagement in collective self-understanding and action. The second principle upon which the formation of such spaces is based is that they are created by groups that may seek to go beyond their own interaction to act as 'counterpublics', in attempts to gain access to broader publics, in order to publicise their own perspectives and contest the 'main' public sphere with the aim of transforming it. These 'counterpublics' thus represent sites of identity formation and sites for developing resources for communicative contestation. Thus, Fraser explains, subaltern counterpublics have a dual character. On the one hand, they function as spaces of withdrawal and regroupment; on the other hand, they also function as bases and training grounds for agitational activities directed toward wider publics. It is precisely in the dialectic between these two functions that their emancipatory potential resides. This dialectic enables subaltern counterpublics partially to offset, although not wholly to eradicate, the unjust participatory privileges enjoyed by members of dominant social groups in stratified societies (Fraser, 1993: 15).

For Fraser, this notion of a public sphere not only paves the way for the democratisation of the dominant public space, but also allows these subaltern counterpublics 'to articulate and defend their interests in the comprehensive public sphere' (Fraser, 1993: 14). For our concerns, however, a particularly interesting aspect of Fraser's account is her suggestion that, to the extent that a proliferation of 'counterpublics' is enabled, this can facilitate an 'expansion' of the 'comprehensive public sphere':

> insofar as these counterpublics emerge in response to exclusions within dominant publics, they help expand discursive space. In principle, assumptions that were previously exempt from contestation will now have to be publicly argued out. In general, the proliferation of subaltern counterpublics means a widening of discursive contestation, a good thing in stratified societies (Fraser, 1993: 15).

Fraser's articulation of the public sphere here could provide a suitable theoretical framework for considering AJE's possible expansion of the 'global public sphere' as a possible 'subaltern counterpublic', since some studies on AJE have regarded the channel as providing a 'voice to the voiceless' (Painter, 2008; El-Nawawy and Powers, 2010).

Fraser argues that a multiplicity of publics presents an appropriate model for democratic participation, enabling a shift from an ideal of consensus stemming from rational critical debate, as proposed by Habermas, to a valorisation of contestation that might progressively support representations of difference. Thus, she argues, '[a] post bourgeois conception would [...] allow us to theorize a range of possible relations among such publics, thereby expanding our capacity to envision democratic possibilities beyond the limits of actually existing democracy' (Fraser, 1993: 26). Through this model of the public sphere, Fraser explicitly contests Habermas's account in terms of its descriptive accuracy and as a critical standard, since the bracketing of status and appeal to rational-critical norms precisely disguises the fact that this model is structured through inequality. In its place, she proposes an alternative modelling of the public sphere through a 'core-periphery' model in which there are dominant and counterpublic spheres, where the latter are both defined against the dominant as well as operate as intra-group spaces of identity and debate. For this reason, this model of the public sphere acknowledges the problem of a sphere of domination and recognises alternative attempts to 'counter-balance' dominant power and transform it.

Overall, Fraser's articulation of the public sphere is a political one, which acknowledges that struggle exists. It is a pluralistic model and hence is a departure from Habermas's rational-critical ideal of a singular public. Fraser's model is based on contesting the terms of public debate, thus acknowledging that there are spaces of micro organisations which aim to transform the wider public sphere. In recent works however, Fraser (2007, 2009) has revisited her depiction of the public sphere in what she describes as a 'postWestphalian' world. This refers to:

a world where notions of inviolable and equal state sovereignty – never actually a reality but often respected as a norm – are breaking down; where states are no longer the sole or even the most important actors in certain areas of international politics; where the "national interest" cannot be defined in one-dimensional terms; where power takes many different forms, both soft and hard; and where the distinction between 'domestic' and 'international' politics is irreversibly blurred (Newman, 2007:3).

In Fraser's revisiting of the public sphere, she argues that most of the literature on the public sphere beginning with Habermas, and including her own, has failed to respond to a more globalising public (Fraser, 2007: 53). Accordingly, she suggests that the emergence of a transnational public sphere might offer new possibilities for democratisation. Recently, Fraser has proposed that '[t]he idea of a "transnational public sphere" is intuitively plausible [. . .] and seems to have purchase on social reality' (Fraser, 2009: 76). As a result, she suggests that a transnational public sphere be seen as a space where 'inclusiveness' is required in a 'postWestphalian' world and stresses the need to address not only 'who' is to emerge as participants but more importantly 'how' (Fraser, 2007: 63), since these two questions are mutually inclusive. This is a fundamental point that is strongly linked to this study's inquiry, as the mere inclusion of AJE within the global news space is not a guarantee that the expansion of the 'global public sphere' is a direct consequence, rather there is a need to uncover whether AJE is *discursively* expanding the 'global public sphere'. A possible outcome to this, as Fraser argues, is that '[t]he challenge, accordingly, is twofold: on the one hand, to create new, transnational public powers; on the other, to make them accountable to new, transnational public spheres' (Fraser, 2007: 65). This is crucial for the development of a global civil society which needs to be as globally representative as possible.

Global civil society is conceptualised 'for operational, descriptive purposes only, as a sphere of ideas, values, institutions, organisations, networks, and individuals located between the family, the state, and the market, and operating beyond the confines of national societies, polities, and economies' (Kaldor, Anheier & Glasius, 2005: 2). This is a definition of global civil society that goes beyond Habermas's national parameters where '[i]n the eighteenth century, civil society referred simply to a society characterised by a social contract; it really referred to a gentlemanly political elite, perhaps elected on limited suffrage, which debated and deliberated about key decisions'

(Kaldor, Anheier & Glasius, 2005: 2). More recently civil society is not just transnational, which is moving beyond borders, but has taken a more global formation, as Mary Kaldor, Helmut Anheier and Marlies Glasius argue:

> What we mean by global civil society is not just civil society that spills over borders and that offers a transnational forum for debate and even confrontation; rather, we are concerned about the ways in which civil society influences the framework of global governance – overlapping global, national and local institutions. Some theorists prefer the term 'transnational' to 'global'. But by 'global' we mean more than just 'beyond borders': we refer to the ways in which globalisation has transformed the issues and problems that we face and the role of civil society in confronting them (Kaldor, Anheier & Glasius, 2005: 2).

Although Fraser critiques Habermas for remaining within the national Westphalian sphere, she points out to the difficulty of conceptualising the public sphere at a global level. Fraser believes that it is

> difficult to associate the notion of legitimate public opinion with communicative arenas in which the interlocutors are not fellow members of a political community, with equal rights to participate in political life. And it is hard to associate the notion of efficacious communicative power with discursive spaces that do not correlate with sovereign states. Thus, it is by no means clear what it means today to speak of 'transnational public spheres' (Fraser, 2007a: 8).

Thus, according to Fraser the concept of interlocutors also changes within the transnational public sphere where 'the interlocutors do not constitute a *demos* or political citizenry', and '[o]ften, too, their communications are neither addressed to a Westphalian state nor relayed through national media' (Fraser, 2007a: 14). For the purpose of this study, I will borrow the concept of 'interlocutors' and apply it to participants within an emergent global media space. Here I apply Fraser's argument that the question at stake is not only 'who' but 'how', where not only is the increase of mediated 'interlocutors' the focus. Fraser explains this process:

> all interlocutors must, in principle, enjoy roughly equal chances to state their views, place issues on the agenda, question the tacit and explicit assumptions of others, switch levels as needed and generally receive a fair hearing. Whereas the inclusiveness condition concerns the question of *who* is authorized to participate in public discussions, the parity condition concerns the question of *how*, in the sense of on what terms, the interlocutors engage with one another (Fraser, 2007a: 20).

The importance of this 'parity condition' goes beyond the need for an increase in the number of representative media channels, but also the need for different voices to attain communicative power in order to participate as 'interlocutors' which practice their discursive right.

The following section looks into the possible emergence of the mediated 'global public sphere' through different positions in the literature, where emerging new satellite news channels – such as AJE – could play a role in the emergence of a civil society which has gone beyond the national and onto to the global arena. Hence, the question which is raised is the extent to which a mediated 'global public sphere' can be realised with the emergence of recent satellite television channels from various parts of the world.

Positions on the Mediated 'Global Public Sphere'

As a result of an increase in the number of representative satellite television news channels, many opinions and positions have emerged in relation to the extent this increase could provide a realisation of a mediated 'global public sphere'. The emergence of 24-hour television news, or 'rolling new channels', has been mapped by Stephen Cushion (2010) through three historical stages which he describes as 'overlapping'. The first stage is the 'coming of age' which is marked by the launch of CNN as the first 24-hour news television satellite channel. It encapsulates the rise and spread of CNN, particularly during the 1990–1991 Gulf war where CNN's coverage was following unfolding events and providing an around the clock reporting (Cushion, 2010: 16–19). The second phase, regarded by Cushion as 'a race for transnational reach and influence', is seen as a period where other channels began to emerge emulating the 24-hour news reporting of CNN (Cushion, 2010: 19). Such channels include the BBC World Service News television channel (its name at the time), which was launched in 1991 (Cushion, 2010: 21). Also, in 1996, AJA, entered the regional scene of 24-hour television news, where it played a significant role in the second Gulf war competing alongside CNN (Cushion, 2010: 22). Following this, the third stage underwent a 'regionalization of the 24-hour television news genre', where not only was there an increase in the number of 24-hour news channels from various parts of the world, but also established news channels, such as CNN, began to branch out to more region-specific markets, such as 'CNN TÜRK, CNN-IBN, CNNj, and CNN Chile' (Cushion, 2010: 24).

The public sphere and also the globalisation of the public sphere have been topics of debate since the 1990s (Johnson, 2006: 100). To a large extent, such debates were stimulated by the emergence of transnational satellite television (Parks, 2003: 75). As Simon Cottle (2009) has suggested, three responses to the question of an existing 'global public sphere' are found in the literature on global news. As he notes, to a considerable extent these debates have been dominated by, on the one hand, a critical and somewhat pessimistic 'global dominance' paradigm and, on the other, a more optimistic (and occasionally celebratory) 'global public sphere'. Cottle defines these two paradigms:

> Studies within the 'global dominanace' paradigm generally work within and update the critical tradition of political economy while those conducted under the 'global public sphere' paradigm represent a more diffuse grouping of recent disciplinary infusions from cultural studies, anthropology, and approaches to the global 'network society' (Cottle, 2009: 28).

In addition to these contesting paradigms, Cottle identifies a third body of work that is 'beginning to qualify the overarching claims of western media dominance and [exhibits] a more theoretically circumspect or cautious stance towards claims of an emergent "global public sphere"' (Cottle, 2009: 32). In order to further contextualise this study in regards to the role of AJE in the 'global public sphere', and to consider what sort of arguments and evidence have been presented in support of these perspectives, this section will discuss these three positions.

Media Imperialism

A first position on the nature of the 'global public sphere' in relation to television satellite news is articulated through the 'political economy' paradigm, which does not support the realisation of a mediated 'global public sphere'. Political economy refers to:

> the interplay between the symbolic and economic dimensions of public communication. It sets out to show how different ways of financing and organizing cultural production have traceable consequences for the range of discourses, representations and communicative resources in the public domain and for the organization of audience access and use (Murdock and Golding, 2005: 60).

The implications of political economy at an international level have been debated mainly through the control of multi-media conglomerates over the global media sector, as conceptualised through the 'media imperialism' thesis. The media imperialism thesis stems from 'cultural imperialism' debates and positions, which Annabelle Sreberny-Mohammadi (1997: 49–50) has mapped through its various articulations: Herbert Schiller's (1976) position on 'cultural imperialism' is summed up as the 'rise of American corporate power and its ideological expansion worldwide via the media'. Cees Hamelink (1983) labels 'cultural imperialism' as 'cultural synchronization' thus relating the process whereby, as a consequence of largely unidirectional 'flows' of culture, an increased homogenisation of practices of cultural production and consumption coincides with an increasingly hierarchical exercise of control over cultural production. John Tomlinson (1991) discusses 'cultural imperialism' through 'Western praxis', such as 'bureaucracy', 'capitalism', and 'urban-industrialism'. Finally, Stuart Hall (1992) focuses on 'cultural imperialism' through the impact of discourse of 'the West on the Rest' by looking into the 'global spread of modernity'.

While these perspectives have focused on the question of whether contemporary global media serve as vehicles of cultural imperialism, others have focused more strongly upon the business and industrial structures of transnational mass media corporations. Such work argues that the presence of multi-media conglomerates taking on the world stage has made the concept of the media operating as a 'global public sphere' and facilitating democratic communication highly improbable. For example, Robert McChesney (2003) argues that the global media industry is best described as characterised by a largely US-based domination. This domination, he suggests, is based on companies wanting to extend their operation to achieve global reach, in order to capitalise on the possibility of increasing financial gain. McChesney views this expansion as one that has been partially enabled by the emergence of communications technology, but argues that the real drive behind this phenomenon is the 'shift to neoliberalism, which means the relaxation or elimination of barriers to commercial exploitation of media, and concentrated media ownership' (McChesney, 2003: 30). Furthermore, he notes, these dominant media are becoming even more concentrated, as a result of smaller players merging with the global media conglomerates (McChesney, 2003: 28–29). The power of these dominant media is used in shaping public opinion by setting the agenda for global consideration, which makes it harder for newcomers to enter the field of global media to challenge the power of conglomerates. Consequently, 'the wealthier and more powerful the corporate media giants have become, the poorer the prospects for participatory democracy' (McChesney, 1999: 2).

One consequence arising from criticisms of 'information inequality', resulting from the effects of the political economy of transnational media production, is a scepticism regarding the possibility of contemporary forms of mass media facilitating a 'global public sphere'. Colin Sparks, for example argues against the consideration of a 'global public sphere' by contrasting the principles of the Habermasian ideal of an accessible, democratic sphere of public representation with the reality of the contemporary media environment (Sparks 2005: 34–48). In current circumstances, he also suggests that a 'global democratic order' is not realistic as a consequence of media being defined and restricted politically, economically or culturally through forms of national regulation. In addition, he argues that as a result of international news networks accessible only to the elite and educated people of the world, the concept of the 'global public sphere' is farfetched. In an earlier, related argument, Sparks also argued that a major factor that lies in the heart of a fully operational mediated global public, is that of language. In order for television channels broadcasting from the 'global South' to enter the global news space and have a significant impact and role to that of dominant media, it is necessary for them to broadcast in English. English is regarded as the prevailing global language, and hence English-based television networks are seen to incorporate what is assumed to be a more global audience. Indeed, Sparks suggests, language is a vital component for the structure of a 'global public sphere'. He argues that without access to a global language, access to a 'global public sphere' is limited. Nonetheless, in his more recent work, Sparks does acknowledge that ' [d]espite its subordination to concentrations of political and economic power, there is no doubt that under some circumstances satellite broadcasting has been able to expand the range of news and discussion quite considerably' (Sparks, 2005: 46). Here, he refers to Al Jazeera Arabic (AJA) as one example. Nevertheless, Sparks remains highly sceptical about the validity of describing the contemporary communicative environment facilitated by mass media as a 'global public sphere'. The force of this scepticism is well demonstrated in the conclusion to his original engagement with the question of whether a 'global public sphere' exists:

> If we need to abandon the term 'global public sphere' as manifestly inadequate to designate what we have been analysing,

then a better one is needed. The one that fits the evidence best is 'imperialist, private sphere'. If this is unfashionable, so be it. At least it is accurate (Sparks, 1998: 122).

Such concerns about the domination of international communication and media imperialism, have a longer history that gained particular attention through calls for a New World Information and Communication Order (NWICO), specifically through the forum of the United Nations Educational, Scientific and Cultural Organisation (UNESCO) in the 1970s and 1980s. As Boyd-Barrett and Thussu (1992) have recounted, the development of the NWICO agenda had its roots in the changing composition of the United Nations (UN), notably with the entrance of newly formed post-colonial states into the UN. Representatives of these states sought to use the UN as a forum of addressing both economic and communication equality, such that UNESCO became both a forum for often strident debate on the problems of information and communication inequality, in criticism that came to focus particularly on the role news agencies played in dominating 'news flows' from the developed to the developing world, while simultaneously tending to reproduce distorted images of the developing world. Dependency critiques directed at the existing communicative order suggested that news agency coverage, reflecting Western and neo-imperial concerns and interests, not only served to dominate international coverage of non-Western events and issues, but tended to relay these perspectives back to the 'South', thereby denying global populations alternative communicative resources for developing perspectives on the problems and issues facing them, and in turn for an adequate and balanced representation in the global public. This provides a background to why it is necessary to evaluate the role of newcomers such as AJE in possibly contesting this dominance. As a result, UNESCO provided a forum within which powerful critiques of what Boyd-Barrett (1997) refers to as 'media imperialism' developed, focusing on underlying problems of both political-economic and information inequality that contributed to a 'communication dependency'. Critical perspectives identified numerous underlying problems in the political economy underpinning news flows: the tendency of news agencies, and international reporting more generally, to develop news agendas relevant to the home markets that provided their primary source of revenue; and to use these secure financial bases to maintain 'economies of scale', allowing them to undercut potential agencies to secure an effective oligopoly among the major news agencies, such as Reuters, Associated Press (AP) and Agence France-Presse (AFP). At the same time, the problems facing the developing world in breaking this dominance, such as the spiralling cost of communications infrastructure, the lack of strong domestic markets and the inability to compete with international rivals were cited as entrenched problems demanding international policy action. This led to the establishment in 1979 of the International Commission for the Study of Communication Problems chaired by Sean MacBride, and the production of what came to be widely referred to as the MacBride report in 1980, which provided both a defining statement of the concerns underpinning the NWICO agenda, and a series of recommendations as to how communication inequality might be addressed. Stating that there were 'various reasons, both inside developing countries and on the international scene, which prevent media from counteracting the blanketing effect of the one-way flow' (UNESCO, 1980), the MacBride report also highlights issues of lack of funding and limited opportunities for the development of professional talent for media in the developing world to exist, survive, and compete in an aggressive international market place.

As noted above, a key target of criticism in the MacBride report focused on global news agencies as commercial platforms of transnational political communication. News agencies are considered to be the global agenda setters (Thussu, 2000: 151; Clausen, 2003: 17), and since there are a few dominant global news providers in operation, news agencies have the power to dictate what is newsworthy and what is not. Boyd-Barrett and Rantanen argue that international news agencies have been representative of Western geo-political and economic interests:

Inevitably, Western agency services were found to be very imbalanced, giving content preference to the region to which a

given news wire was directed and then to the news of the
United States and Western Europe; giving greatest weight to
political, economic, and military news, news of international
affair, and international sport (Boyd-Barrett and Rantanen,
2004: 34).

To address such a reality, initiatives in the MacBride report also called
for the cooperation between news agencies and the developing world;
promoting the interest from Western media to publish news from
developing countries; the creation of resource centres in non-Western
parts of the world; and the establishment of regional news agencies
(UNESCO, 1980: 143). However, while the NWICO debates
did support some significant initiatives, particularly through the
establishment of the non-aligned news agency pool and the provision
of support to news agencies in the developing world, the NWICO
agenda was strongly opposed by Western member-states, specifically
the US and Britain who were, and arguably remain, the two most
dominant players in international news markets. The departure of the
US and Britain from UNESCO in 1985, resulted in a significant
decline in both revenue and influence for UNESCO (Thussu 2000,
Carlsson 2003).

Recently, the NWICO debates have been revisited through the
International Telecommunications Union (ITU) which sponsored a
set of talks on international communications at the World Summit
on the Information Society (WSIS). In a comparative study between
both debates resulting from NWICO and WSIS, Victor Pickard
argues that:

it is clear that the space for debate at WSIS has narrowed.
For example, it is noteworthy that many related issues,
including transborder data flow, cultural diversity, and the
regulation of news content, are no longer on the table for
discussion at WSIS. This shrinkage of debate indicates a
further dominance of neoliberal logic and transnational
corporations, thus altering national priorities and discursive
boundaries (Pickard, 2007: 136).

The call for a balance in information flows in relation to the MacBride report has been particularly evident in the Arab world; which is an area that has witnessed a plethora of foreign media broadcasting to the region without any Arab media broadcasting directly back to the Western world. As Tourya Guaaybess notes, as a response to the NWICO report, 'Arab countries also called for a re-balancing of communication flows and for autonomous means of communication: they had to provide the world with their own image of themselves' (Guaaybess, 2002: 9). Under the auspices of the Arab League, a meeting for the Council of Arab Ministers of Information occurred in September 1967 in order to consider 'for the first time the possibility of using satellite technology' (Guaaybess, 2002: 9). According to Guaaybess, '[t]his technology could serve, in accordance with the discourses, as a counterweight to the information of non-Arab countries' (Guaaybess, 2002: 9). As a result of this the pan-Arab system of satellites, ARABSAT, was set up (Guaaybess, 2002: 9). Thus, the NWICO debates could be regarded as having an effect on the development of transnational media in the Arab world.

More recent work focusing on the political economy of international news has suggested that, in many respects, the major concerns regarding imbalances in media production and flows of news remain valid. Thussu, for example, notes that international news agencies are responsible for the distribution of more than 80 per cent of global news (Thussu, 2004: 51; 2000: 152), with the exception of recent usage of social media sourcing by journalists – evident during the Arab Spring uprisings – this monopoly remains unchanged. This serves to emphasise that the continuing media power agencies hold in constructing global reality becomes apparent since news agencies do not only source news, but also play a significant role in setting global news agendas. In addition, these major news agencies have developed a long-term alliance with dominant global news retailers. For example, although CNNI sources most of its own news due to its global wide spread, it subscribes to all three news agencies. BBC, on the other hand, has always subscribed to Reuters but also sources its news from AP. As for News Corporation, it has always sourced much of its news from

Reuters (Paterson, 1997: 147–148). These solid relationships between news agencies and news media, Paterson suggests, both reflect and enforce the monopoly of the 'Western' domination of global news. Boyd-Barrett has also revisited his original definition of media imperialism as 'the process whereby the ownership, structure, distribution or content of the media in any one country are singly or together subject to substantial external pressures from the media interests of any other country or countries without proportionate reciprocation of influence by the country so affected' (quoted in Boyd-Barrett, 1998: 165–166). Boyd-Barrett acknowledges some of the criticisms that have been levelled against critiques of approaches that depend on a centre-periphery model of global communication, particularly in light of the development of multiple centres of media communication. He also highlights that his original definition did not adequately take into account audience reception. It is naive to presume that audiences in developing countries, who themselves face the political, economic, and social reality of their everyday life, would take at face value the dominant media's side of the story. A parallel set of issues have been raised by Peter Golding and Phil Harris (1997) and Rantanen (2005: 77–78) who, like Boyd-Barrett, combine a sympathetic critique of the media imperialism tradition with an effort to argue for its continued merits. Similarly, Boyd-Barrett defends the continuing merits of his concept, which he argues lie in its ability to accommodate a variety of media practices in addition to a range of dimensions and levels of media activity, while attending to the continuing problem of hierarchies of communication capacity and inequalities in media flows.

However, while it may still be possible to defend such a critical focus, developments in the field of satellite television news do appear to have undermined some of the assumptions upon which debates around satellite news have been structured, particularly that of a simple 'one-way flow' from international centres of news production to 'peripheries' in the global 'South' (Rantanen, 2005: 78–79; Cunningham, Jacka and Sinclair, 1998). In their study of the more recent ecology of satellite television, Mugdha Rai and Simon Cottle (2007) have highlighted what appear as simultaneous trends towards both domination and proliferation in the transnational television industry. Thus, in their mapping of global satellite television news networks, Rai and Cottle emphasise both the status of CNNI, BBC World, and Fox News as the key Western players with the largest reach, but note the simultaneous emergence of multiple regional and local satellite news channels as a significant feature of the news ecology, not least because it may well be these are channels that audiences prefer to watch. Yet despite an increase in satellite television 'representation' where various parts of the world are broadcasting their news via transnational satellite television, a few of these channels have gained global reach. For example, in North America and Canada, they note that there is an increase in local media broadcasting from local regions which cover localised topics. As for South America, there are fewer channels than North America but there are some prominent satellite television stations such as the Brazilian Globonews, Argentina's Todo Noticias and Venezuela's Telesur Network. Europe, on the other hand, has more channels, although Euronews and Sky News are the only satellite television channels broadcasting within the region. Throughout South Asia, Rai and Cottle (2007) note a domination of Indian media with particular interest in 24-hour news. However, in East Asia Taiwan is a media centre alongside significant broadcasters such as China's English CCTV-9, Hong Kong's Phoenix News, and Singapore's Channel News Asia. As for the Middle East, Al Jazeera is expanding its reach through its English-language satellite channel AJE. Nevertheless, despite the global emergence of satellite television channels Rai and Cottle also highlight some silences — many of which remain today. For example within the Oceania region Australia, New Zealand and the Pacific Islands, Sky News Australia is dominant, and mainly includes content from its parent company Sky News UK. In such cases, Boyd-Barrett's definition of 'media imperialism' as a concept that may apply to regional dominations may very well be apt. In a starker example, Africa is also a region of marked silence since it does not have any of its own satellite channels, although there is a strong Western and Asian presence there. However, generally speaking, these developments may suggest areas that the media imperialism thesis neglects and which needs to be taken into consideration when conceptualising the field of satellite news. Indeed, these developments are frequently presented in support of a second position, which regards the field as becoming increasingly representative.

'Global Public Sphere' as Emergent

Where, as we have seen, a focus on 'media imperialism' tends to be associated with scepticism towards the idea that a 'global public sphere' exists, the second position relating to this question is a positive one, which regards the field of global news as increasingly characterised by a multiplicity of voices, refuting the media imperialism thesis. Such arguments point to the manner in which satellite television channels from various parts of the world have generally been on the increase – thus resulting, it is claimed, in the emergence of an unprecedented number of cultural spaces and variety of discourses. Attempts by non-Anglo and non-American global broadcasters to enter the global media space in recent years have increased, attempting to counter-balance the one-way flow of information addressed in the section above. Recently, the emergence of global media broadcasting from such news networks has added a fresh perspective to the debates on global news flows. Prasun Sonwalkar argues that '[t]he rise of international news channels from countries of the South, carrying a variety of Southern perspectives, is probably the single biggest change in the global media landscape over the last decade. Such a development challenges the old paradigm of media imperialism and is broadly considered to be a good thing' (Sonwalkar, 2004: 111). According to James Lull, this rise of international news channels has led to the reversal of imperialist patterns of global flows, and what can now be better articulated as 'diverse and complex' flows (Lull, 2007: 11–12). Lull argues that 'trends in global media influence today often decentralize, challenge, and realign traditional sectors of institutional authority in the spheres of politics and culture, just as early printing technology did centuries ago' (Lull, 2007: 11). Thus, Lull suggests, today's global media, facilitated by technological change, serve to further open up spaces for democratic dialogue and representation.

In another study on global media developments, David Demers (2002: 156–173) also identifies an increase and differentiation of global media as a key trend that will affect mass media industries in the twenty-first century. This trend is a result of a continued growth in the number of global televised services, with more specialised channels identifying the needs of particular audience groups. Hence, Demers suggests, global media corporations will become less powerful due to a decline in their ability to control information markets. This is due to the increase in the number of media outlets where consumers will have many choices that suit their needs, such that no single media corporation will be able to dominate the market as state-run and private networks from Western countries did in the previous century. Thus, Demers argues:

> To be sure, news media organizations like CNN (an AOL Time Warner company) will continue to reach hundreds of millions and probably billions of people, and their audiences will expand. But needs for information will continue to become more specialized, and many alternative news channels and outlets will emerge to serve those specialized needs (Demers, 2002: 163).

Thus, according to Demers, large-scale media will still generate more revenue and have a more competitive edge even in specialised areas due to their capital resources, yet smaller specialised media will also be on the increase since larger media do not always find it profitable to offer similar specialised services (Demers, 2002: 164).

While Demers thus suggests that market trends need not undermine the possibility of increased levels of representation, Ingrid Volkmer suggests that these are resulting in an emergent 'global space' of representation, facilitating a 'representative public zone' that has 'cross-cultural, cross-societal, and cross-national implications'. This 'global space', Volkmer argues, 'can be considered as a multi-discursive political space' (Volkmer, 1999: 122–123). Volkmer (2002) argues that during live crisis reporting, 'authentic angles' are necessary, and regards new satellite television channels that provide such angles as 'micro-spheres' by offering new perspectives to both national and international dominant broadcasters. She states in relation to these new entrants to the field that '[i]t can be argued that in future they will increasingly create counter-flows to mainstream news coverage – internationally and domestically – and create "micro-spheres" in an extra-societal, "global public sphere"' (Volkmer, 2002: 241).

Volkmer discusses these 'micro-spheres' through the example of Al Jazeera's role in the September 11 attacks and on the war in Iraq, which could be regarded as a successful counter-flow at an international level. Volkmer also brings in the example of the Thai Global Network, run by the Royal Thai Army Television, which reaches 144 countries worldwide including the expat Thai community in the US (Volkmer, 2002: 241–242). Other examples of 'extra-societal' television channels are those which broadcast from outside the national country targeting local audiences, as in the case of the National Iranian Television (NITV) which broadcasts from Los Angeles to 'influence politics in Iran by an in-flow of critical information delivered by satellite' (Volkmer, 2002: 242). Such examples, Volkmer argues, 'reveal a new dualism of supra- and sub-national journalism, creating political communities within nations' (Volkmer, 2002: 242). According to Volkmer, the process of globalisation marks a change to the meaning of 'public' as understood in previous debates, since the public sphere has expanded beyond the limits of the national onto a global context (Volkmer, 1999: 122). She argues that 'the global communication sphere, involving fragmented satellite television targeting specific audiences worldwide and the internet, provides not so much 'global' communication but particular globalized communication' (Volkmer 2007: 58). Indeed, she stresses, contemporary global media flows are not based on 'place', but rather on 'space':

Globalization of today's media flows are not arranged in a 'place'-based nexus, distributing a signal within one geographical and cultural territory from a to b through identifiable (and licenced!) channels and gatekeepers. It is rather an advanced 'space'-based nexus, providing new discourse spheres which could provide conceptual frameworks for information sovereignty (Volkmer, 2007: 61).

It is important to note that Volkmer claims that information sovereignty 'could' occur, not that it already has. What is notable about this position is that, rather than simply inverting the position of, for example, Sparks – who presents a model of a public sphere as a critical standard against which what he regards as the inadequacies of contemporary global media – Volkmer suggests that a 'global public sphere' is increasingly *being* realised, rather than claiming that it already has been realised.

Whilst, in this way, Volkmer positions the field of satellite television news as increasingly becoming representative, Brian McNair's 'chaos' perspective leads him to dismiss the 'media imperialism' thesis altogether. McNair argues that there is a shift from global media control to that of an anarchic circulation where a variety of players are operating. He suggests that smaller media players now have a place to function independently, as 'more media, moving more information further and faster means a more chaotic communication environment, with corresponding implications for the acquisition and management of power in society' (McNair, 2006: xx). McNair uses the term 'cultural chaos' to signify 'various disruptions accompanying the emergence of the globalised news culture of the twenty-first century' (McNair, 2006: xii). The concept of chaos is linked with the scientific concept of a system that is determined by non-linear laws where chaotic systems are regarded as 'fundamentally unpredictable' (McNair, 2006: xii). Thus, according to McNair, 'media organisations are the agents of cultural chaos' (McNair, 2006: xx). In comments that are particularly relevant to our concerns he suggests that the concept of cultural imperialism is losing its strength with the current growth of 'real-time' satellite news, if it is not already 'wholly redundant' (McNair, 2006: 116). Real-time satellite news, he suggests, refers to a *flow* medium characteristic of around the clock live news coverage, as opposed to a 'medium of record'. The phenomenon of real-time satellite news, McNair argues, forms a 'global public, accessing a common news source' (McNair, 2006: 109). McNair's argument, thus, is that there is a shift from the 'manufactured' and 'constructed' thesis of news production which produces predictable outcomes in terms of content, to that of a 'chaos model' which is not structured as there is no one fixed agenda setter due to the multiplicity of actors:

> The chaos model thus stresses unpredictability of outcome in media production processes, a consequent uncertainty around the quantity and quality of information flow, the importance of feedback loops, and enhanced volatility in the management of both communication and power. From this perspective news is not manufactured (neither, therefore, is consent), nor is it 'constructed'. Nor does it just happen. It *emerges* from the interacting elements of the communication environment which prevails in a given media space (McNair, 2006: 49).

News production, according to the chaos model is a natural process that is a result of a multi-dimensional news environment that is becoming more influential than a one-way flow of domination. According to McNair, this leads to a 'loss of control, dilution of authority, and expanded opportunity for disruption of elite power' (McNair, 2006: 50). McNair's chaos model, however, describes the field of international news as an arena that cannot be understood in terms of structured practice. While the global news environment is undergoing major changes, describing it as chaotic is problematic since satellite television news channels operate through relations of power, which means the discourses they produce are a systematic result of this. McNair's argument has, therefore, raised some concerns among scholars, as a consequence of his tendency to present a stark binary between apparently simplistic 'control' models, and his own 'chaos' thesis. Cottle (2009), for example, suggests that the field of positions relating to contemporary global journalism is more complex than McNair's argument allows. McNair's characterisation of the 'control paradigm', Cottle suggests, 'runs the risk of perpetuating its own reductionism in that it tends to caricature a field that is far more theoretically differentiated and a little more sophisticated than suggested' (Cottle, 2009: 41). However, the sphere of media power is somewhat dismissed in McNair's thesis, which is problematic since media power is inherently part of the fabric of the media environment. Yvonne Jewkes (2004) has also criticised McNair's earlier work for its reproduction of a pure liberal-pluralist perspective on media that pays insufficient attention to the way market structures tend to place limits upon representation. The pluralist perspective, she suggests 'could be said to be limited by its sheer idealism' since it dismisses the reality of 'media ownership or control', or the 'profit-oriented nature of much media output' (Jewkes, 2004: 22). Indeed, she argues, '[e]ven 24 hour rolling news services on cable and satellite, such as CNN, are restricted by the news values to which they have to conform and by the pressures of having to succeed in a commercial environment' (Jewkes, 2004: 23). Rather than a chaotic model in which media markets increasingly become more attuned to the needs and interests of all viewers, Jewkes suggests that 'the increasingly commercialised character of media institutions results in tried and tested formulae, with an entertainment bias, aimed at a lowest common denominator audience who are easily identifiable and potentially lucrative audiences for advertisers' (Jewkes, 2004: 23). Jewkes' argument highlights the commercial reality of the media as an industry, which the chaos model tends to dismiss. Although the recent increase in the number of satellite television channels is an optimistic development from the media imperialism thesis, it is necessary not to dismiss a centre-periphery model as argued by Fraser in relation to a main public space. Jewkes' position echoes an earlier argument made by Curran, who argues that one critical issue with the liberal approaches of the public sphere, in general, is that they do not 'take adequate account of the way in which power is exercised through capitalist and patriarchal structures' (Curran, 1991: 29).

It could be argued that an increase in the number of satellite news channels, such as the inclusion of AJE into the global media space, have enabled the emergence of an increasingly representative 'global public sphere'. This is a position which tends to characterise media globalisation itself as an 'opening' process, and is therefore associated with a process of 'democratisation'. Here the emergence of global satellite television news channels, such as AJE, are seen as examples of mediated 'interlocutors' (Fraser, 2007a) that could possibly play a role in the promotion of a global civil society. In this way, promoters of the emergence of a 'global public sphere' offer a brighter and more optimistic view to that of the reproduction of 'Western' media dominance through 'media imperialism'. However the entry of new players from previously unrepresented parts of the world into the main public sphere in its own token might not be a guarantee of addressing the imbalance created by the corporate domination of global media. While acknowledging that developments in global media markets may indeed offer some democratising possibilities, a third set of arguments have sought to present a more cautious analysis than such bold predictions regarding the realisation of a democratised global space allow.

Domination and Resistance

The third position – which has been recently emerging in the literature on global news flows and which this book identifies with – recognises that there is some validity in both sets of positions discussed above. This position does not, by contrast to strong versions of the media imperialism argument (such as that presented by Sparks) involve a total denial that a mediated 'global public sphere' exists, but neither does it seek to claim that an *ideal* 'global public sphere', equating to the realisation of democratised communicative relations, is increasingly being realised. This third position encapsulates both current debates on global news flows, and which leads to considering the existing tension of what I label as *domination and resistance* within the global news ecology – a practice that this study identifies with AJE's role within the 'global public sphere' as I shall go on to discuss in this book.

The second position discussed in the previous section, supports an increasingly representative mediated 'global public sphere' in the number of satellite channels, however as Fraser argues at the start of this chapter, it is not only about *who* is included in the 'global public sphere' but also *how* through a variety of discursive representations.

Hence, the question of the 'global public sphere' should not only be argued through the increase in the number of satellite news channels but also through the nature of their discursive input within the global news arena. Cottle and Rai highlight the 'inherent complexity' within the communicative architecture of global televised news, stressing that the field 'should no longer be simply ignored or collapsed in reductionist fashion under theoretical positions of "global dominance" or "global public sphere" – though both can summon empirical findings in partial support of their respective claims' (Cottle and Rai, 2008: 176). While Cottle and Rai's approach assists us in gaining an understanding of the complexity of the field of international news, there is a need to conceptualise the actual discursive nature of the mediated 'global public sphere' in relation to how mediated 'interlocutors' (Fraser, 2007a) serve to expand the mediated global news space. Despite the persistence of hierarchies and inequalities of the field of satellite news, however, Rai and Cottle argue that the field is 'dynamic and rapidly expanding', such that 'traditional configurations of "centre" and "periphery" seem to be increasingly shifting' (Rai and Cottle, 2007: 61). This trend seems to be missed by research that has tended to focus on dominant broadcasters, thus excluding other satellite television activity that is emerging:

> When we examine the full range of contemporary 24/7 channels now available around the world we necessarily move beyond a Western-centric view and from this vantage point can better conceptualize, and theoretically reflect on, the changing satellite news landscape, its complex stratifications and contemporary reconfiguration. In short, there is a regional, transnational and global complexity here that demands increased recognition and theorization (Rai and Cottle, 2007: 53).

While it stops short of identifying globalisation with democratisation, the position represented by Rai and Cottle's work suggests that a view of globalisation that treats it as merely a continuous 'flow' of Western dominance underestimates the complex relations that exist between global, regional, national and local developments. An illustration of this can be found in the example of US based satellite news television conglomerates which broadcast to national markets within Latin America. These global companies found that domestic media channels in countries such as Brazil's Globo and Mexico's Televisa were in fact dominant in the region; and that programming models and formats that were successful in the US did not attract South American audiences, which led to the need for these American channels to adapt their content to local tastes thus resulting in forms of localisation and in some cases to forms of hybridisation (Straubhaar and Duarte, 2005: 216–219). Joseph D. Straubhaar and Luiz G. Duarte explain the process and its implications as follows:

> [G]lobal firms tend to have approached regional markets by making the most minimal adaptations they felt they could get away with. Since most audiences are more interested in cultural proximity, however, the result of minimal adaptation is also a minimal audience. In order to try to attract larger audiences, this study argues that global firms had to make increasingly larger adaptations to local cultures (Straubhaar and Duarte, 2005: 224–225).

Hence, 'rather than classic imperialism', they argue that there is 'asymmetrical interdependence', whilst at the same time 'national and regional media firms [are] gaining considerable market power against the giant global firms that entered the local markets via satellite and cable television technologies' (Straubhaar and Duarte, 2005: 218). Such examples expose a weakness in the media imperialism thesis, and highlight the power of local markets and local audiences. A similar position on developments in global media is also argued by Terry Flew (2007), who confirms the continued hegemonic power of large-scale, multinational media corporations, yet points to complex and changing relations between local, national and global media:

> Attempts to extend market dominance beyond national home bases by global media corporations will continue to have to

engage in an ongoing manner with key national players, as well as with the diversity and heterogeneity of local and national cultures, and the available evidence thus far suggests that they are at some disadvantage in doing so, in contrast to our usual assumptions about the ubiquitous power of multinational corporations (Flew: 2007: 215).

On this basis, Flew disputes claims regarding the domination of global media corporations within national markets. Instead, he suggests that local and national media regulation has experienced a shift from nation states' control of media flows within their boundaries, to a position where they serve as 'enabling states' which promote media flows 'from the national level to that of the sub-national and the supra-national' (Flew, 2007: 212–213). In light of this, Flew also suggests that a centre-peripheral model cannot capture the complexity of media flows in conditions of globalisation.

The relation between transnational and national publics has also been highlighted by Bernhard Peters (2010) who believes that within the transnational public sphere, international reporting tends to frame and select discourses in relation to national affiliations. This is evident when '[i]mportant national differences [. . .] appear to have persisted not only in respect of agenda-setting and the distribution of interest, but also in respect of the typical interpretative frames underlying international reporting' (Peters, 2010: 241). At the same time, he points out that news circulating within some national publics (some more than others) is becoming increasingly international, which he argues could be due to 'the size of the country' in addition to 'the density of external economic links' (Peters, 2010: 241). So although Peters stresses the role of communication technologies in 'driving' the 'internationalisation processes', he also emphasises that content produced through international news reporting is a product of the inter-relationship between transnational and national publics.

It is notable that this third perspective, rather than representing a purely 'optimistic' or 'pessimistic' account, offers a more grounded perspective on the relationship between globalisation and developments in the media field. For example, rather than offering a general perspective on the 'global public sphere' debate, Cottle and Rai's (2007) work, discussed above, is instead attentive to differences across satellite news at global, regional, and local levels, and the relationship between such differences and political, economic and cultural factors that influence news output. In this respect, their work builds upon Cottle's (2003) earlier work, which has sought to shift away from overarching and generalised perspectives on news output, to consider the complex range of influences that serve to shape it. Thus, Cottle (2003) provides a useful overview of factors that have been regarded as shaping news production where he explains that the understanding of media production has generally been conducted through two lens, the first of which is the *instrumental* explanation' which focuses on the 'proprietorial intervention and the instrumental pursuit of ruling interests and/or political allegiances'; the second position is 'structural determination' which understands media production through focusing on 'agency, and maintains that media personnel, whatever their position within organisational hierarchies, are compelled by impinging determinants – economic, technological, ideological – to reproduce media forms and content unconsciously and routinely in predetermined ways' (Cottle, 2003: 5). Cottle points out that there are various other forces that come to play during the process of news making which are often overlooked. These forces constitute a 'middle ground' where there is a need to 'consider the possibility of the complex interplay between these different factors and the dynamics that come into play within particular fields of production' (Cottle, 2003: 4–5).

This perspective on news being subject to complex and contingent, rather than purely determinist, forms of influence has strongly informed the field of journalism studies. Michael Schudson, for example, refers to four approaches to understanding news production through 'the economic organization of news', 'the political context of news-making', 'the social organization of newswork', and 'cultural approaches' (Schudson, 1995). The first two perspectives are related to the economy and the state, the third perspective reflects the social aspect of the 'construction of ideology' within the news network, and the final perspective 'emphasizes the constraining force of broad cultural traditions and symbolic systems' (Schudson, 1995: 174). Other studies have also considered the different factors that affect the making of news. Rodney Benson (2004), for example, looks into the influences that shape mediated political discourse and provides a 'recategorization of key influences' on the 'journalistic field'. Benson argues that culture needs to be considered contextually and in relation to other influences, such as political and economic power. Accordingly, Benson suggests classifying the influences of political news coverage through '(a) commercial or economic, (b) political, and (c) the inter-organizational field of journalism' (Benson, 2004: 280). He does this by looking into 'individual organizational and journalistic factors' which affect 'the broader organizational and professional field' (Benson, 2004: 280). Here Benson argues the need for historical and cultural considerations to be factored into these three major influences in order to 'explain the origin and solidity (or lack thereof) of the journalistic field's relation to political and economic power' (Benson, 2004: 280). In relation to the economic factor or 'commercial pressures', Benson lists these as 'concentration of ownership, level and intensity of competition, profit pressures related to type of ownership, and type of funding', in addition to 'government or political restraints' (Benson, 2004: 281). When considering other influences within the 'interorganizational field of journalism', Benson discusses a set of factors which shape news discourses, where he highlights the complex and multilayered nature of the journalistic field, arguing that the field cannot be diminished into one or two forces of power, an important point to keep in mind when considering the nature of the mediated 'global public sphere'. According to Benson, a variety of factors include 'reporter-source relationship'; the 'ideological allegiances' between cross-national journalistic fields which give way to various journalistic models; in addition to journalists' professional and individual values (Benson, 2004: 275–292). These overall factors not only influence the discourses projected, but also determine the nature of the news organisation itself.

In another study exploring the influences which affect media content, Pamela J. Shoemaker and Stephen D. Reese (1996) outline what they label as the 'hierarchical model'. Starting from the bottom of the 'hierarchical model' is what they position as the 'individual level' which could be understood as the 'influences on content from individual media workers'. At this level, the 'factors that are *intrinsic to* the communication worker' are considered, such as their education, personal background and professional history (Shoemaker and Reese, 1996: 64). Additionally, other considerations include their value and belief systems, and how they perceive their roles (Shoemaker and Reese, 1996: 64). This is followed by the 'media routines level', thus referring to the 'influence of media routines' which is associated with 'an organizational perspective on the mass media' (Shoemaker and Reese, 1996: 107). Then the 'organization level', meaning the 'organizational influences on content', which look into the nature of 'roles performed, the way they are structured, the policies flowing through that structure, and the methods used to enforce those policies' (Shoemaker and Reese, 1996: 172–173). The 'extramedia level' follows which ties with 'influences on content from outside of media organizations' and that incorporate 'sources of the information', 'revenue sources', the 'economic environment', 'technology', and 'other social institutions' which include governments and businesses (Shoemaker and Reese, 1996: 175). At the top of the 'hierarchical model' – and which includes all the above influences – is the 'ideological level' which refers to as the 'influence of ideology'. This influence is concerned with relations of power since 'media transmission of ideology works as it does by drawing on familiar cultural themes that resonate with audiences' (Shoemaker and Reese, 1996: 222). Overall, these factors play a significant role in the construction of news discourses, particularly when it comes to considering global journalistic forces within a global news space (Reese, 2001). In this book I refer to *forces of discursive media power* which this study has identified as factors that influence and shape media power within the global news sphere, and which will be discussed in the final chapter.

It is by reference to such complex levels of influence that authors within this third paradigm have suggested that, simultaneous to trends towards conglomeration in the media field, a simultaneous trend towards pluralisation and the opening up of new platforms of representation may also be significant. Although a fully-representative mediated 'global public sphere' has not been realised – as we saw from Rai and Cottle's (2007) mapping of global satellite news channels around the world – there is no doubt that there is an increase in the number of media channels from different parts of the world that have gone beyond the restrictions resulting from the media imperialism model. While it would be naive to completely ignore the existence of dominant global media conglomerates, it might yet be the case that the current global media space is undergoing a transitory phase. Mizuko Ito argues that '[i]f networked media ecologies are maturing and becoming more established in our everyday lives, we are also still clearly in a moment of transition' (Ito, 2008: 3). What is significant about the arguments in this third position in relation to debates surrounding the mediated 'global public sphere', is that while they do not suggest that these trends produce an inexorable trend towards an increasingly democratic 'global public sphere', they suggest that neither can contemporary globalisation simply be equated with an extension of imperial domination from 'centres' of global media production. In particular, the tendency to seek out new markets through the diversification of media products, new technological potentials for media production and dissemination, and the development of emergent forms of competition within local and global media markets, may provide opportunities for new forms of representation to emerge. At this stage, there is a need to acknowledge the emergence of additional new 'voices' in mainstream media as emphasised by Couldry, who argues that:

New political acts (not just moments of spectacle) will need to be *recognized* in the news coverage of mainstream large-scale media; any expansion of the cooperative process within democratic politics must be represented in media coverage. Media institutions are very good at voicing 'counter-democracy', but it is unclear whether they are good at reflecting new forms of political cooperation or 'ordinary democracy' [...] Perhaps media's role in politics needs to be reframed as part of this process (Couldry, 2010: 148).

Such reframing is necessary for the creation of a more representative and global mediated space where more 'voices' are represented. Couldry maps various understandings of 'voice' but focuses on 'voice as a value', which can be articulated through 'people's practice of giving an account, implicitly or explicitly, of the world within which they act' (Couldry, 2010: 7). Here Couldry relates 'the value of voice' to Fraser's framework which refers to the need for 'particular groups' to be heard after the 'long-term exclusion from effective voice' (Couldry, 2010: 3). Couldry argues that '[s]paces for voice are therefore *inherently* spaces of power; their link to power does not just derive from institutions such as government seeking to manage them [. . .] it must be places in a sociological context which is always, in part, a political context' (Couldry, 2010: 130). Similarly, Fraser states that 'a public sphere is conceived as a vehicle for marshalling public opinion as a political force. Mobilising the considered sense of civil society' (Fraser, 2007a: 7). Through these arguments, there is a common consensus on the need for the mediated 'global public sphere' to be more representative and for it play a more politically active role in mobilising civil society. Power and public communication is being re-phrased where 'voice' (Couldry, 2010) and 'interlocutors' (Fraser, 2007a) are pointing towards news forms of power in transnational publics.

For the purposes of this research the problem we are concerned with is whether the rise of AJE, as an offshoot of its parent channel AJA, can be understood as a development that, in Fraser's terms, serves as a means through which what was previously a 'subaltern counterpublic' can increasingly contest dominant assumptions, 'expand discursive space' and facilitate a 'widening of discursive contestation' (Fraser, 1993: 15). To answer this question and provide a more grounded basis for assessing the extent to which AJE might be understood to comprise an 'alternative' source in this sense, I shall now turn to consider debates surrounding the question of AJE acting as an alternative voice through historic and scholarly debates, in addition to an account of AJE's self-presentation.

CHAPTER 2

AL JAZEERA AS 'ALTERNATIVE'?

Al Jazeera English (AJE) has been promoting itself as a channel that offers an 'alternative' source of news, which is firmly based on Al Jazeera Arabic's (AJA) reputation of alternative broadcasting. Within the Arab public sphere, AJA has been known for explicitly attempting to present information to the 'global public sphere' from an Arab or Muslim perspective, with the aim to challenge what it sees as dominant perspectives coming out of global news networks such as CNNI or BBC World (Nisbet and Myers, 2010: 351). Accordingly, scholars have placed AJA as playing an alternative role in radicalising debate (Volkmer, 2007; McNair, 2006) and considered AJE as giving a 'voice to the voiceless' (Painter, 2008; El-Nawawy and Powers, 2010). Al Jazeera, as a news organisation, is regarded by many as an example of a satellite television station which began broadcasting in the form of alternative media as a 'micro-sphere' contesting state power (Volkmer, 2002), but then developed beyond the national to the global. Al Jazeera's aim as a news organisation, in general therefore, has always been to provide a distinct voice on social and political issues and debates that have not previously been voiced within the mediated Arab and 'global public spheres'.

This chapter uncovers a detected 'tension' in AJE's self-promotion as a channel that is attempting to balance itself between the Arab and global spheres. It outlines the debates surrounding AJE's launch, as a relatively recent broadcaster in the global news space and as an offshoot to AJA, through existing scholarly debates and through AJE's own articulation of its operation. This chapter will start by historically outlining Al Jazeera's 'alternative' role from within the Arab and global spheres, followed by a mapping of how AJE promoted itself prior to its launch in context with this historical role. To critically consider this, however, it is necessary to briefly define what is meant by 'alternative' media, despite the fact that AJE does not bear much similarity to the amateur and grassroots operations that are typically associated with this sector.

Debates about 'alternative media' have come to question approaches which seek to define alternative media through a set of qualifying traits. Alternative media have evolved through time. The emergence of 'oppositional journalism' was noted during the period of Habermas's bourgeois publics, which is a form of journalism Atton and Hamilton equivalently label today as 'alternative journalism'. They argue that, 'oppositional journalism' was, during this period, regarded as 'the coin of the realm for legitimate public discourse and debate, regardless of the purpose or cause', whereas today bourgeois journalism is regarded as 'the main target for challenge' (Atton and Hamilton, 2008: 17). Thus, Atton and Hamilton suggest that the majority of journalism in the eighteenth century was not yet an integral element in modern power in the way they later became. By contrast, today alternative journalism often tends to regard mainstream journalism, and in turn 'media power' (Couldry, 2003), as a key target for challenge. Christian Fuchs argues that the role of alternative media 'should not only be understood as alternative media practices, but also as critical media that question dominative society (Fuchs, 2010: 174).

Despite the common awareness regarding the difficulty of defining alternative media, various attempts have been made to provide a comprehensive categorisation. One definition attempts to position alternative media through mainstream media practice. One popular term that is introduced by Downing is 'radical', as he points out the issues related to the term 'alternative':

> [I]t must be acknowledged that to speak simply of alternative media is almost oxymoronic. Everything, at some point, is

alternative to something else [...] To some extent, the extra designation radical helps firm up the definition of alternative media, but even here, we need to make some preliminary qualifications (Downing, 2001: ix).

Downing explains that radical media may be seen as negative or constructive depending on the viewpoint, and in some circumstances, radical media could also encompass ethnic minority, religious or community media (Downing, 2001: x). Yet the term 'radical' can be restrictive and in turn exclude other media practices that are not necessarily radical but still produce counter-discourses. Accordingly, the label 'alternative', as Chris Atton argues, offers a more general application, because 'radical' 'encourages a definition that is primarily concerned with (often revolutionary) social change' (Atton, 2002: 9). He firmly advocates that alternative media, as a general application, provide information and interpretations of the world that would not be found anywhere else (Atton, 2002: 12). This is similar to Fraser's conceptualisation of 'subaltern counterpublics' which provide discourses that are not dominant and that aim to challenge hegemonic discourses. Similarly, Couldry associates the term 'radical' with political practices of the media in which they 'express an "alternative vision" to hegemonic views of the world' (Couldry, 2002: 25). For that reason he also prefers to use the term 'alternative' media since it provides a more generic label to that of 'radical' (Couldry, 2002: 25). Therefore such definitions label alternative media practice as a counter-practice to mainstream media.

A second definition addresses the notion of alternative media practice in relation to various users of alternative media and their representation. For example Rodriguez argues for using the term 'citizens" media instead of 'alternative', as the term 'alternative' leads to relating this kind of media to a binary style of thinking in opposition to the mainstream (Rodriguez, 2001: 20). Rodriguez develops the term 'citizens" media by linking it to forms of citizenship practice and empowerment that are strongly tied with the everyday lives of citizens (Couldry and Curran, 2003: 7; Atton, 2003: 267). However, among alternative media activists, practitioners and academics there has been some concern voiced over the use of the title *citizen* as it overlooks problems and issues related to the status of citizenship, thus excluding those who are regarded as 'non-members' of the state.

In line with such labelling, some popular definitions have centred on 'community' or 'grassroots' media. Although such definitions constitute a representative articulation, they do nonetheless offer limitations to such practices that need to grow and reach beyond whom they represent, in order to follow Fraser's model of finding a position in the main public sphere. Yet again, in regards to these terms, they do reflect more of what they exclude, which is mainstream media, than what they signify (Downing, 2001: 40). Defining alternative media 'against' mainstream media has been common practice, as many regard alternative media as distinctive from dominant mainstream media. Tony Harcup argues that '[t]here is a long and continuing tradition of alternative media being produced to challenge the discourse(s) of mainstream media' (Harcup, 2003: 356). Here, the limited characteristics of mainstream media discourse stem from their habit of 'maximizing audiences by appealing to safe, conventional formulas, and "alternative" foregoing the comfortable, depoliticizing formulas to advocate programs of social change' (Hamilton, 2000: 357). Accordingly, James Hamilton suggests that mainstream media adopt traditional customary frameworks in order to attract audiences, whereas alternative takes on original contesting frameworks in an attempt to create social change. Downing concurs with this now, arguing that 'the term alternative media is reserved for certain categories [...] politically dissident media that offer radical alternatives to mainstream debate' (Downing, 1995: 240). Hence, the idea that a defining characteristic of alternative media challenging discourses and debates produced by mainstream media is shared by many scholars.

I have argued elsewhere (see ElGhul-Bebawi, 2009) that the most suitable definition of alternative media is that of 'contesting media power' (Couldry, 2003), as it is no longer viable to categorise alternative media as only 'citizens', 'grassroots' or 'community' media especially at a time when social media are increasingly playing a dominant and commercial role. Therefore, the categorisation of whether a media channel is 'alternative' or not needs to be conducted through the network's discursive content rather than its organisational parameters. Accordingly, it is the discursive content that determines a news network's degree of 'challenge', and not how a media channel promotes itself. Nonetheless, Al Jazeera's transition from the Arab to the global spheres instigated a notable campaign by the news network to promote AJE's agenda and position as a global operator, and which is worthy of study.

In order to assess how AJE promoted itself, it is first necessary to outline Al Jazeera's journey from the Arab public sphere and understand something of the history of AJA and how this served to position the new channel. It is not within the scope of this book to address the wealth of debates and perspectives that have been generated around AJA. Rather, I am concerned to engage with this literature insofar as it is important to provide some of the debates that have emerged around AJE as an emergent global alternative broadcaster. It is also necessary to note that the developments outlined in this section on the formation of the mediated Arab public sphere are discussed through the role of broadcast media with specific focus on television, and not on early media such as the press or recent media such as the internet. This book is concerned with satellite television broadcasting, and thus the outlining of the mediated Arab public space will be conducted through this lens. Moreover, although foreign media have been operating within the Arab public sphere from the initial stages of media activity in the region, this chapter will still consider the influence of such media but remains specific to Arab self-representation.

The Mediated Arab Public Sphere

In this following section, the mediated Arab public sphere will be historically discussed through three different transitional stages: the first is the *national model* where Arab media mainly operated within the national boundaries of the state; the second model is a phase where the Arab public sphere took on the *transnational model* thus marking the spread of satellite television channels which began transmitting across national borders, a period through which AJA played a noted role; and the third model marks the post-September 11 period when discourses central to the Arab public sphere started to gain an international presence in news discourse through AJA, thus marking the start of the integration of the Arab sphere into the *global*. These three stages will be discussed in detail.

The National Model

Al Jazeera, as an organisation, began operation in the Arab region at a time when Arab television stations were restricted to broadcasting within the boundaries of the state, and were primarily government owned and controlled. Audiences in the Arab world, therefore, could only access television stations that were controlled by the nation state they were residing in. There were, however, a few cases where television signals from neighbouring countries could filter through as a result of the short distances between the two states (Rugh, 2004: 201). Such examples include the Arabic-language service for the Israeli television channel that broadcast into Jordan, and which provided nightly news on the Arab/Israeli conflict and the internal politics involved. Generally, Arab populations received their news from their state broadcasters who used the media 'as a political tool to mobilize the masses and propagate the official line' (Rugh, 2004: 184). The government played a role in the manufacturing of news where 'Ministries of Information provided the editors with guidance on political programming [. . .] instructing editors to ignore sensitive issues rather than to exploit certain themes for their propaganda value' (Rugh, 2004: 193–194). News, therefore, to a large extent functioned as a platform for the state. This was evident in evening news bulletins which 'consistently presented evidence of the latest achievements of the government, and extol the virtues of the top personalities' (Rugh, 2004: 193). As Naomi Sakr argues, this impeded television from acting as a 'fourth estate' within the Arab world, since '[b]roadcasters cannot be relied upon to act as a watchdog on government if the way they are regulated makes them subject to government control' (Sakr, 2007: 18).

Consequently, Arab audiences did not develop a trusting relationship with their media and slowly became a sceptical audience. Hugh Miles notes that

Arabs learned to despise and distrust everything they heard, read or saw in the media. All the media came to be regarded, quite rightly, as appendages of the government, which only ever echoed, never investigated or criticized, what their leaders said (Miles, 2005: 25).

This resulted in mistrust between many Arab people and their national media, which was further augmented during periods of major crises such as the initial wars with Israel, and at a later stage the Gulf war in 1991, as Miles illustrates:

Whenever Arabs began to turn back to their state media, for example in times of war, their trust would be disastrously betrayed. The most famous instance of this was during the Arab-Israeli war of 1967, when Arabs everywhere were glued to the Sawt al-Arab radio station founded by Gamal Abdel Nasser, President of the United Arab Republic (Egypt). The beloved announcer Ahmad Said, a household name in the Middle East, declared that the Arab armies had crushed the Israeli army and that Israeli planes were 'falling from the skies like flies'. The rest of the Arab media went on to repeat this message until a week later, when Arabs found out from foreign sources that they had, in fact, been utterly defeated. Arab trust in the media was shattered. Since then the media has done little to win it back (Miles, 2005: 26).

Media from outside the nation state were also received with some degree of caution. Incoming transnational radio stations broadcasting in Arabic from Western origins gained presence within the Arab public sphere such as the BBC Arabic News Service, Radio Monte Carlo, and Voice of America which was particularly seen as a propaganda service for the US. Miles notes that '[a]lthough these stations were extremely popular and offered a higher standard of news than anything produced domestically, they were Western and so still subject to some suspicion' (Miles, 2005: 25). Global news television networks broadcasting from outside the Arab world, such as CNN, were introduced at a later stage but were only available to the few who could afford it. During the Gulf war in 1991, CNN was the sole foreign broadcaster broadcasting from within Iraq, thus marking the first modern-day televised coverage of war in 'real time'. The power of this central role played by CNN was noted by Arab audiences and governments alike, and consequently 'the strategic possibilities of satellite television were reconsidered' (Miles, 2005: 26). This in turn led to the emergence of 24-hour news satellite channels within the Arab world, as will be discussed below.

Overall, the mediated Arab public sphere during the national phase was a very restricted model. Arab audiences relied on discourses which circulated within Arab streets and in private gatherings. With such a tightly controlled mediated Arab space, the advent of transnational satellite television was seen by many as a phenomenal opening up of the communicative space. In view of that, the satellite television industry in the Arab world witnessed a tremendous growth throughout the region.

The Transnational Model

Satellite technology was introduced to the Arab region with the launch of ARABSAT satellite in 1985. However, it was only a couple of years later that this technology was put to use (Miles, 2005: 26). After a long period of state media control, where Arab audiences could only get their news from government-controlled media, there was a need for them to be exposed to other perspectives and other agendas. Equally, private enterprises were also awaiting opportunities to launch their own satellite channels. With the spread of satellite television technology in the Arab world, this meant that both private investors and Arab audiences alike could make use of this new technology. Culturally, this meant that the Arab world had opened up to each other, and Arab audiences could further be exposed to different Arab publics. Politically, this meant that multiple political views reached the Arab living room from national, transregional and transnational channels.

Issues of access were generally not a problem, as the mushrooming of satellite dishes can be seen across rooftops and on balconies in the Arab world. Marc Lynch (2006) provides an insightful account of the mediated Arab public sphere and its evolution from state control to the opening up of satellite television stations. He notes that the transnational mediated Arab space has resulted in connecting Arab and Muslim countries together, in addition to including and engaging the Arab diaspora around the world (Lynch, 2006: 52). Accordingly, transnational Arab media share similar issues, beliefs and identity that form the basis of their mediated discourses (Lynch, 2006: 52). Despite the cultural and linguistic differences across the Arab world, this transnational commonality ultimately formed an Arab public sphere that had a shared collective set of discourses. These discourses mainly centred on counter-discourses in response to a perceived Western hegemony. As a mediator of these discourses, AJA was seen as a key player in the development of a mediated Arab public sphere. Lynch explains the effect of AJA on the Arab public sphere where he suggests that 'Al Jazeera rose to prominence by giving voice to public opinion rather than directly attempting to mobilize or lead it' (Lynch, 2006: 37).

The timing of AJA's launch within the Arab world is important to note, since AJA came at a time when Arab audiences were sceptical about the coverage of state Arab media during the Iraq war in 1991, and the disappointment that loomed as a result of misleading coverage. AJA brought to the average Arab home a variety of debates, topics and views that were not openly circulated amongst society. This sparked noise and drew attention to the newly established channel. Hafez al-Mirazi, chief of AJA bureau in Washington said at the time: 'Al-Jazeera has a margin of freedom that no other Arab channel enjoys. Our motto is "the view and the other point of view"' (quoted in El-Nawawy and Iskandar, 2002: 34), and as the Qatari Foreign Minister stated 'one opinion and the other opinion' (Rinnawi, 2006: 55). The emergence of AJA was therefore seen by the Arab audience as a channel that allowed for open discussions within the Arab public sphere. Hence, the reason behind AJA's regional recognition fundamentally lies in its practice. As Zayani and Sahraoui put it:

> By Arab standards, the very existence of Al Jazeera is revolutionary. Al Jazeera has brought noteworthy innovations to Arab broadcasting and reporting, airing hard-hitting programs, bold and uncensored news coverage, passionate political debates, and on-the-ground reporting. The network introduced new programs and institutionalized certain program formats, broadcasting many of its shows live on the air, soliciting viewer participation and taking heed of the voices of ordinary people [. . .] Al Jazeera is particularly noteworthy for breaking boundaries and building a reputation for being outspoken on issues traditionally deemed as sensitive or taboo in the Arab world. It has been hailed as a beacon of free press, a bold initiative in journalism, and a revolutionary force among Arab media long constrained by state control (Zayani and Sahraoui, 2007, 23–24).

Despite many Arab governments regarding AJA as problematic due to its bold reporting, which is usually a direct attack on these governments, the above account of AJA has been prevalent and explains why AJA shot to fame within the Arab public sphere. Although this description of AJA is rather celebratory in tone, this is quite comprehensible in light of the historical context prior to its emergence.

Nonetheless, there have been concerns regarding the extent to which AJA serves as a democratic platform for alternative views. El-Nawawy and Iskandar point out to what they call 'media schizophrenia', because although AJA is critical of Arab governments, its coverage does not include much on its domestic politics (El-Nawawy and Iskandar, 2002: 83). The response to this from AJA is that Qatar is a small nation with not much news to report on. Another reason is that Qatar does not play a key role in Middle

Eastern politics. (El-Nawawy and Iskandar, 2002: 84). Despite this, El-Nawawy and Iskandar point out that although

> Al-Jazeera began with a $140 million loan underwritten by the Qatari government, supplemented by a similar contribution that would cover operations until November 2001 (when the loans came due) [. . .] During the five-year loan period "the channel has been studying other means of financing" (El-Nawawy and Iskandar, 2002: 89).

Sakr argues that AJA did not intend on being a radical voice, but was broadcasting with the aim of being a 'model of pluralistic reporting' (Sakr, 2007a: 115). AJA, she suggests, 'was not primarily conceived as a source of counter-hegemonic contra-flow. It was based on a widely accepted model of pluralistic reporting espoused by, among others, the BBC' (Sakr, 2007a: 115). This 'model of pluralistic reporting' was new to the Arab world, which made it radical to the region; and as an Arab broadcaster, then became radical to the world. Nevertheless, Sakr argues that AJA 'fits media theory's concept of contra-flow as it applies to a reversal of old imperialist imbalances' (Sakr, 2007a: 104). In spite of this, she suggests that this performance as a 'contra-flow' may be only partial, since 'contra-flow in its full sense would seem to imply not just reversed or alternative media flows, but a flow that is also counter-hegemonic' (Sakr, 2007a: 105). In other words, AJA's flow from the Arab region to the global media platform alone does not label it 'alternative' or 'radical', as there is a need to establish that the discourses it presents do in fact challenge those of dominant media. So, from Sakr's viewpoint, in order to test a station's claim to alternative practice, there is a need to consider the station's news agenda. She argues that '[f]or such assertions about a challenge to be justified, it needs to be demonstrated that Al-Jazeera consciously pursued a distinctive news agenda and that this agenda was knowingly obstructed in countries where globally dominant media are based' (Sakr, 2007a: 106).

According to Sakr (2005), AJA, within the Arab public sphere, became a prominent and mainstream voice. In a study on AJA's position in Arab media, Sakr conducts a mapping of Arab transnational television and demonstrates how, despite many television channels attempting to emulate AJA in various Arab countries, they have not been that successful due to restrictions, be it political, social or economic – though mainly political. She investigates the impact AJA had on Arab satellite channels and gives an overview of the emergence of satellite television in the Arab world, highlighting various cases of satellite television stations that have attempted to emulate AJA as an alternative medium but failed as a result of financial restrictions, policy regulations, or government pressure. However, she argues, restrictions through media laws on editorial freedom of journalists working in transnational television have also had an effect. In this respect, in contrast to romantic perspectives on the libratory force of satellite television, she notes that 'transnational television companies, although transmitting through space, have their headquarters on the ground' (Sakr, 2005: 86). By contrast, she argues, AJA succeeded in providing perspectives different to those provided by other Arab broadcasters.

The Global Model

From the time that AJA began providing the world with messages from Osama bin Laden, AJA became globally visible. Through this example, Lynch describes the Arab public as emerging as a 'counterpublic, self-consciously and intentionally challenging the dominant narrative and terms of discourse within American and global media' (Lynch, 2006: 57). Lynch notes that the '[t]he Arab public sphere has long identified itself as a subordinate, dominated counterpublic, struggling against Western hegemony and tenaciously resisting pressure to conform from all sides' (Lynch, 2006: 56). Within the Arab public sphere this in turn 'generated a rich Arabist hidden transcript, an alternative, coherent, widely shared interpretation of political structures and relations that could not be openly aired or translated into practice because of the realities of weakness and subordination' (Lynch, 2006: 57). However, these discourses were not known to Western or global publics due to mainly linguistic differences, which in turn enclaved the Arab public sphere from global audiences. The intensity to which such discourses were entrenched within the Arab public sphere, as a result of Arab media constantly driving them, resulted in an Arab audience who may have thought that such discourses were common to Western and global publics too. Yet Lynch points out that such discourses were 'alien' to Western discourses, and draws on the example of mediated messages that were broadcast to American audiences by Osama bin Laden:

> When Americans heard Osama bin Laden for the first time, most found his rhetoric literally incomprehensible – laden with symbolism, history, assumptions, and references that resonated within Islamic public spheres but were totally alien to the American public [. . .] Statements and political rhetoric that made perfect sense in one public sphere, tapping into well established motifs and languages, seemed literally incomprehensible in others (Lynch, 2006: 58).

This example highlights a 'gap' between the regional Arab public sphere and the 'global public sphere'. The relationship between these two publics is not based on the familiarity with the discourses of the other, but rather is an oppositional relationship stripped of the understanding of the other. As Lynch puts it, '[t]he relationship between a subordinate Arab public sphere and an American-dominated international public sphere suggests an analysis in terms of domination and resistance rather than dialogue' (Lynch, 2006: 57). This lack of dialogue is problematic, and hence the need for Arab representation within the 'global public sphere' was not only vital for the development of a democratic space but also for the understanding of the other.

In relation to AJA's role within the 'global public sphere', although seen by some as democratising the Arab public sphere, AJA's crisis reporting has always been problematic from certain Western and particularly American perspectives. AJA has been regarded as instigating hatred, encouraging resistance and promoting 'anti-American sentiments throughout the Middle East' (El-Nawawy and Iskandar, 2002: 180–181). For example, AJA's coverage of the Afghanistan war included images of Afghani children casualties, which were reported by AJA correspondents on the ground as being caused by American bombing (El-Nawawy and Iskandar, 2002: 180). This and many other examples continued not only in relation to the Afghani crisis, but also in relation to the Iraq war and the Israeli/Palestinian crisis. However, it was after the attacks of September 11, when bin Laden began using AJA as a platform for sending messages to the Western world (particularly the US administration) that the station gained global fame. Mixed with its reputation of being defiant to Arab regimes, AJA was categorised as an 'alternative' channel that was perceived as not only controversial but, by many, as dangerously radical.

By contrast, others regard AJA as a positive contributor to democratisation at both local and global levels. Volkmer considers AJA as a channel that plays an alternative role both within the Arab sphere and the 'global public sphere'. She argues that AJA's role within the Arab sphere 'can be seen as attempting to counterbalance conservative Arab state channels within this particular microsphere' (Volkmer, 2007: 68). Consequently, Volkmer argues that this counter-balance entered the 'global public sphere' during periods where there was a global focus on the Arab world. She states that 'it can be claimed that these microsphere spaces gain particular prominence in times of world crisis, when these fragmented "flows" gain intensity and provide "authentic" feedback and information within the global public space' (Volkmer, 2007: 68). McNair, similarly, suggests that AJA offered news discourses within the Arab world that were also not previously incorporated within the Arab public sphere. He argues that AJA 'communicates a new model of mediated politics which can only be beneficial to the broader project of democratisation' (McNair, 2006: 164). At a global level, McNair argues that AJA has challenged 'the traditional global dominance of Anglo-American news providers, not just in their local areas but in Europe, America and Australia' (McNair, 2006: 116–117). So both at an Arab and global level, McNair sees that AJA has and continues to play a role in radicalising debate. However McNair does point out that AJA has had to soften its tone slightly for the following reasons:

> Whether through choice (to improve the quality of the brand in the global news market) or necessity (to stay on the air), Al Jazeera's managers have subsequently modified their approach to coverage of the Middle East. Since 2004, as part of a process of international consolidation and expansion [...] Al Jazeera has sought to strengthen its ability to report diverse views in a manner consistent with liberal pluralistic norms, and has softened its support for Islamic fundamentalism by giving greater access to moderate Islamic as well as western views and voices (McNair, 2006: 114).

While for some such moves are positioned as a problematic 'watering down' of AJA's potential to offer a genuinely distinctive voice, for McNair such adaptation is presented as an unproblematic process of democratisation. This adaptation to the Western world, he concludes, has positioned AJA as an 'Arab media service with a global orientation' (McNair, 2006: 114).

By contrast, Iskandar (2006) has suggested that the question of how 'radical' AJA is depends upon both its historical context and its reception. Indeed, Iskandar argues that AJA is now regarded as 'mainstream' in the Arab world, whereas in the West audiences regard it as alternative. As he puts it: '[c]urrently, the station is in an ambivalent position "*vis-à-vis*" its regional and global audiences – in some instances serving as the sole voice of discursive dissent and in others acting as the major mainstream broadcaster in the Arab world' (Iskandar, 2006). El-Nawawy and Iskandar position AJE as practicing 'contextual objectivity', which is 'the necessity of television and media to present stories in a fashion that is both somewhat impartial yet sensitive to local sensibilities' (El-Nawawy and Iskandar, 2002a: 54). What this means is that a media channel will try to balance itself between the notion of objective reporting and its target audience. This term according to El-Nawawy and Iskandar is intentionally oxymoronic since it reflects the tension which attempts to produce an objective message to a culturally sensitive audience (El-Nawawy and Iskandar, 2002a: 3).

Overall, the debates surrounding the nature of AJA's reporting vary in regards to the degree to which AJA is seen to act as an alternative platform. Nonetheless, the consideration of AJA as an alternative broadcaster within scholarly debates sets the background for considering whether AJE can be regarded as an alternative news network, given that it has a different target audience to cater for. The following section will provide a critical mapping of the various opinions within media commentary and AJE's own self-promotion which fluctuate between views which promote it as an extension of AJA's radical reporting and those who fear that AJE would be a 'watered down' version of AJA.

AJE's Self Promotion

Prior to the launch of AJE in November 2006, a lot of public speculation was generated about the sort of contribution an English-language news network that stemmed from AJA might make to the field of global news. Given the reputation AJA had as a 'radical' channel, particularly following various public criticisms that had been made of it that had positioned it as the voice of terrorism (El Amrani, 2006), an expectation arose that AJE might constitute both a source of radical perspectives that were previously unavailable to global audiences via high-profile, transnational satellite broadcasting, and serve to constitute an 'Arab voice' that would present distinctive – and perhaps oppositional – perspectives on international news events. Much of this expectation was fuelled by AJE's own marketing and self-presentation, which sought to draw on AJA's reputation to position itself as offering a unique and important perspective. This section will provide a discussion based on the results of a mapping of newspaper articles, online commentary, staff interviews, and through AJE's own broadcasts on how it promoted itself as a global broadcaster during its beginning stages.

AJE began broadcasting on 15 November 2006 from Doha, Qatar. By 2009 it was estimated to be 'accessible in over 120 million households worldwide' (Powers and El-Nawawy, 2009: 269), and had also increased its online presence (Powers and El-Nawawy, 2009: 269) through platforms such as YouTube. AJE is regarded as the first English news network satellite television channel that broadcasts from the Arab world at a global level. Although AJE caters for a Western audience rather than an Arab audience, it still wanted to be perceived as an authority in the Middle East (Barkho, 2007: 23). Originally, the channel was intended to be called Al Jazeera International (AJI), but changed its name to Al Jazeera English (AJE). AJE shares its logo with AJA. This logo – which is an artistic calligraphy of the word Al Jazeera in Arabic – has become a visual global brand and one which AJE uses as a reinforcement and extension of that brand.

AJA has been keen to promote itself, and has therefore been 'actively engaged in publicity through press releases, communiqués, statements, interviews, and the like' (Zayani, 2008: 215). As part of its broadcasts, AJE featured a quote that promotes its aims in these terms: 'Al Jazeera is about perspective. The Arab language channel was about presenting a new perspective. Now we're doing it in English' (Broadcast 18 March 2007). In this way, AJE explicitly positioned itself both as an extension of AJA and as a media channel that offers a different and alternative perspective. Accordingly, AJE attempted to promote itself as providing discourses that are not otherwise addressed in English-language news media. Other examples of such positioning noted from recorded study data were the use of slogans such as: 'Reversing the North to South flow of information', 'Setting the News Agenda', 'Every Angle / Every Side', and 'All the News / All the Time'. Such language is reminiscent of the historic NWICO debates which address the media imbalance within global news flows, discussed in the previous chapter.

With the establishment of the English satellite channel, AJE's audience reach was to include the global non-Arab speaking audience who were curious to know the discourses that AJA's coverage incorporated. As an English-speaking channel, the target reach for AJE was the English-speaking audience across the world which included '40 million households in Europe, Africa and Southeast

Asia' (BBC News, 22 June 2007). Hence, AJE not only aimed to target the Western English-speaking world, but also targeted those who sought a non-traditional perspective of the news as indicated in newspaper coverage surrounding its launch. For example, in Britain's *Independent* it was reported that:

> The target audience of AJI is the millions of inhabitants – Muslim or otherwise – of regions such as south-east Asia, many of whom speak English as a second language. It's a potential audience of one billion, who have different priorities to traditional television news viewers (Jenkin, 2006).

This quote clearly identifies AJE's audience as those who seek non-mainstream news and who would want to watch AJE because it would offer different coverage of news than that of the established broadcasters. Alan Fisher, AJE's Europe correspondent, commented that:

> We aim to look at how real people are affected. I really don't want to attack the BBC or CNN, but they are aimed squarely at businessmen sitting in hotel rooms. We've got a different agenda, to reach the audience you'll find on the street (Jenkin, 2006).

In this sense, AJE promoted itself prior to its launch as a broadcaster explicitly in contrast to both BBC and CNNI, citing differences in its coverage, its agenda and its target audience. Despite AJE here identifying the people on the street as a primary target audience, it also expressed interest in reaching influential segments of the world population. Nigel Parsons, Managing Director of AJE at the time of its launch, stated:

> We're trying to reach educated decision-makers and young people [. . .] We would love to have an audience that regards us as their first source of balanced and impartial news. Beyond that, we'll always be an interesting alternative source (Dana, 2005).

These quotes, which pinpoint AJE's target audience, are indicative of the station wanting to position itself an 'alternative source' that broadcasts to a younger audience 'on the street', whilst at the same time adopting a role similar to that of a dominant mainstream broadcaster where the 'decision makers' go to it as a primary source of news, made all the more valuable for its provision of an 'alternative perspective'. This tension surrounding the dual role that AJE presents itself taking within the 'global public sphere' may reflect its ambiguous position as a quasi-commercial broadcaster that seeks to market its 'alternative' credentials.

Fisher also suggests that AJE staff continuously distinguish the channel from Western broadcasters, and contends that AJE 'will take a global view, rather than looking at things from a purely Western perspective' (quoted in Jenkin, 2006). This is echoed by Rebecca Lipkin, an AJE London-based executive producer, who says she joined AJE because it would allow her to cover stories on parts of the world that were not included by other networks (Dana, 2005).

However, AJE not only presents itself as providing a global perspective, but also specifically an Arab one. According to Parsons, AJE is 'the first news channel based in the Middle East to bring news back to the West [which sets] a different news agenda' (People's Daily Online, 3 August 2007). By saying 'bring news back', here Parsons explicitly references the concept of 'contra-flow' used in the NWICO debates. He offers an example to illustrate his point: 'When our rivals covered the verdict of the Saddam trial, they went back to London and Washington for the reaction of Middle East experts. Our experts are Arabs in the Middle East' (Donnelly, 2005). In this way AJE provides a more authentically comprehensive form than its rivals: 'From its unique position within the Arab and Muslim World, looking outwards, al-Jazeera International reports inclusively, examining all perspectives of a story and providing a fresh 360-degree approach to news coverage' (quoted in Hiro, 2006). Here AJE claims comprehensiveness by including an alternative perspective. Echoing this statement, when asked to sum up the station mission, Parsons commented:

Basically, we will be the first global news channel based in the Middle East looking outwards, and we think we will be bringing in a completely fresh, 360 degree perspective to the news and thereby revolutionize viewer choice, reverse the flow of information; perhaps present the news from a different perspective, and therefore be a conduit to greater understanding between different peoples and different cultures (Pintak, 2005).

It is interesting to note here that Parsons is using the terminology found in previous policy and academic debates when saying 'reverse the flow of information', an indication that the station explicitly positions itself as a 'contra-flow' to Western media, a continuation of AJA's 'contra-flow' to the West (see Azran, 2004). The notion of contra-flow is not only professed by AJE itself, but may also appeal to potential viewers who expect AJE to be a broadcaster that will provide a different view to that of other mainstream broadcasters, such as CNNI and BBC World. For example, one blogger offered his/her perspective on why they would want to watch AJE even if it were blocked from transmitting via cable in the US: 'If freedom of choice is constrained are we not constraining the very foundation of democracy and the right to freedom of expression? I, for one, would be interested to hear a different perspective than CNN or Fox "News" or MSNBC' (CCR, 2006). This blog was posted on 24 July 2006, prior to the launch of AJE. It suggests not only certain expectations that AJE will provide an alternative perspective, but also positions AJE as extending, perhaps even fulfilling, the ideal of a global 'free market of ideas'.

Despite these positive expectations by AJE, others were fearful that AJE might not become a counter-balance since it might compromise its position in exchange for a Western audience appeal. This could be necessary since AJA is often viewed by Western conservatives as operating as a terrorist channel. El Amrani notes that 'negative views of Al Jazeera continue to proliferate, especially among conservatives' and that '[t]hese views are echoed frequently on conservative blogs, including some of the most popular websites in America, which routinely refer to Al Jazeera as "Jihad TV" or "The

Osama Channel"' (El Amrani, 2006). Some feared that this would lead AJE to take a softer approach than its parent channel in order not to alienate its potential audience, and consequently lead to a loss of its Arab identity:

> Even before its first broadcast [. . .] the new channel has been the subject of vicious rumors and scathing criticism from Arab employees who fear that Al-Jazeera's in-your-face journalism will be watered down for Western audiences. They worry that Al-Jazeera International, or AJI as it's known here, will undo a decade-long struggle to build a brand with a distinctly Arab identity. (Allam, 2006).

The concern expressed here is that AJE might become merely a niche global broadcaster which focuses on Arab issues without directly becoming a counter-power to other global dominant networks, as a result of its quest for popularity amongst Western audiences. Such a scenario was regarded as a negative influence on the original Arabic station, as '[m]any worry that the new channel could hurt the reputation of its predecessor if it strays too far from the original Al Jazeera's editorial line' (El Amrani, 2006). As Lawrence Pintak (2006) demonstrates through interviews with AJE staff, such concerns were also shared by AJE journalists. One of the responses he received highlighted that AJE might not follow the stance of AJA which became known for defying American and authoritarian rule in parts of the Arab world. As one source commented, '[m]any are afraid it will not reflect the honest channel these people have sacrificed for' (Pintak, 2006).

Yet, although AJE is seen as a continuation of AJA, and as we have seen it has been promoting itself through AJA's 'radical identity', AJE has also been careful not to promote itself as a direct 'translation' of AJA to ensure it could break into the global market which has been suspicious of AJA (Zayani, 2008: 219). For example, AJE has been found to employ different terminology to that of AJA. The reason behind this, according to Mostefa Souag, Director of Al-Jazeera Centre at the time, is that AJA is a reflection of the Arab culture, whereas AJE is broadcasting to a different audience and will therefore need to adjust its language usage accordingly (Barkho, 2007: 23). By the same token, it is important to note that BBC and CNN do not compromise their terminology when directing their news towards an Arab audience in their Arabic news websites, a point that Leon Barkho makes as a result of a comparative discourse analysis between AJE, BBC and CNN online news (Barkho, 2007: 23).

The above views found in media commentary reflect two positions surrounding the launch of the new channel: one, prominent in promotional materials and media interviews, positions AJE as an 'alternative' channel that, simultaneously, represents AJE as both authentically Arab positions and provides a 'comprehensive' '360 degree' perspective on both global and Arab news events; a second more critical perspective, expressed fears that the station would merely be a 'watered down' version of AJA, and that such 'watering down' might also compromise its parent channel's identity and operation. These views suggest that it might be difficult for a station like AJE to strike a balance between its Arab identity that it has historically built through AJA and its concern to appeal to audiences, despite it promoting itself as an alternative broadcaster.

A recent study on AJE's audiences in Asia, the Middle East, Europe and North America by El-Nawawy and Shawn Powers (2010) suggests that 'AJE's model of journalism offers an alternative to today's mode of news journalism that continues to encourage sterotypical and counter-productive attitudes towards cultural "Others"' (El-Nawawy and Powers, 2010: 62). In this study, El-Nawawy and Powers argue that AJE provides what they label as 'conciliatory media' which is a concept that they define 'as any news media that work to meet a number of criteria [. . .] when covering issues of collective importance [. . .] that is more conducive to cooperation, negotiation and reconciliation' (El-Nawawy and Powers, 2010: 69). In this sense, they suggest that AJE is playing a role in offering 'global understanding of events' (El-Nawawy and Powers, 2010: 62). Based on this study, they argue that AJE journalists and news producers focus on offering 'a voice to the voiceless', which in turn leads to 'mediatized recognition' of the 'underrepresented' (El-Nawawy and Powers, 2010: 62). In addition, the audiences studied found AJE to serve a 'conciliatory function' thus becoming less arbitrary towards other groups (El-Nawawy and Powers, 2010: 62). El-Nawawy conducted this research by interviewing AJE journalists and news staff, in addition to surveying viewers who were familiar with AJE's programming and dependant on it as a source of information. For this reason they targetted participants from Islamic centres and mosques where 72 per cent of the overall respondents were Muslims (El-Nawawy and Powers, 2010: 74). This sample of research resulted in responses that were in favour of AJE's 'conciliatory' role, measured against a scale that draws on the work of Cottle (2006). In addition, El-Nawawy and Powers present some instances where AJE reporting has been particularly praised, both through recognition in media awards and through anecdotal praise, as well as some evidence that AJE audiences appear less 'dogmatic' the longer they watch the channel. What is questionable, however, is the degree to which such phenomena can be attributed to AJE. For example, it may be that the channel's good reputation, and the strong claims that it does serve to provide 'a voice to the voiceless', themselves lead viewers to regard it as a 'good performer' against these criteria. There is no comparison to how viewers of other media, such as CNNI or BBC World, are viewed by their audience. Likewise, there is no particular grounds for suggesting that a lack of dogmatism among AJE viewers is caused by the channel's output: rather, it may be the case that the channel's marketing and reputation serve to attract viewers that already tend to be more open-minded. In this way, the evidence presented by El-Nawawy and Powers, while interesting, reproduces a problem that has been raised by 'effects' based research: that it can only present correlations of phenomena with data as facts, while suggestions of a causal relationship can only be inferred (Corner, 2000). While a 'conciliatory' role is, no doubt, one which would assist in narrowing the gap which has historically developed between the Arab and global spheres, an analysis of AJE's textual output would further clarify whether AJE is playing an actual 'conciliatory' role.

By contrast to El-Nawawy and Powers' work, another recent study by Tine Ustad Figenschou provides a quantitative content analysis of AJE's news and finds that AJE potentially does provide a contra-flow, but at the same time incorporates 'political bias' which characterises mainstream international news (Figenschou, 2010: 85). Figenschou's study focused on the question of whether AJE covered news stories from the global 'South' more than stories from the 'North', in addition to the question of whether the channel offers a 'voice to the voiceless' by interviewing grassroots sources (Figenschou, 2010: 91). In relation to the first question, this study found that AJE stories were more 'from and about the global South than the global North', but were also more focused on 'political crisis and violent/armed conflict' than stories from the global 'North' which were mainly on economic and legal issues (Figenschou, 2010: 98). In relation to the second question on the use of grassroots sources, Figenschou found that AJE uses elite sources which represent the 'establishment', whereas only a few news stories were sourced from 'ordinary people' (Figenschou, 2010: 98–99). As Figenschou concludes, 'Al-Jazeera English's choice of sources illustrates how the channel's alternative agenda of providing a grassroots perspective on world events comes into conflict with the elite orientation in international news journalism' (Figenschou, 2010: 103). Figenschou's study is therefore useful for this study by providing a quantitative counterpoint to this study's qualitative approach, which further investigates the question on whether AJE acts as an alternative voice.

In addition to these studies, James Painter (2008) conducted a study on AJE by comparing its news coverage to BBC World and CNNI, and concludes that AJE can be regarded as a form of 'counter-hegemonic' news source. Painter's study suggests that AJE offers more coverage than the BBC World and CNNI on events from the developing countries with focus on the Middle East, whilst also seeking reactions to international stories from these regions. Painter's study also found that AJE includes news stories from 'under-reported' parts of the world, where there is some evidence to show that AJE provides 'more voices of the voiceless' (Painter, 2008: 44). Painter methodologically conducts this study through a content analysis by noting the amount of time dedicated by each channel to the coverage of particular news stories. While this finding is significant, it may be argued that this does not offer an adequate ground to consider whether this news coverage constitutes a 'counter-hegemonic' practice without a more detailed, qualitative analysis of the nature of AJE's news reporting. In this respect, the discursive analysis conducted in this book may be positioned as an attempt to provide some evidence to consider whether such claims may be sustained if we move from a simple quantitative analysis of content to a consideration of AJE's news discourse.

Overall, AJE has promoted itself as acting as a voice from the Arab world that counter-balances the domination of Western news flows, as exemplified by the global dominance of US and European-based satellite news retailers such as BBC and CNNI. Such claims have been, to a considerable extent, echoed in academic work that, either tacitly or explicitly, served to position AJE's significance in terms of its capacity to provide distinctive and alternative accounts of reality. The broader consideration of debates surrounding alternative media reflect that, rather than having a stable, singular definition, there is considerable variety between the institutional arrangements, funding bases, ideologies and practices among alternative media. For this reason, rather than seek to provide a set of definitions on what alternative media should look like, there is a need to define alternative media in terms of what it does; that is the degree to which it provides alternative discourses which challenge dominant discourses with the aim of understanding and engaging with social reality. As we have seen, there is a detected tension in the way AJE presents itself entering the 'global public sphere'. This tension suggests that it would be difficult to label AJE as purely 'alternative', as this tension is indicative of its transfer from the Arab on to the global spheres and its attempt to balance itself between an Arab audience and a global audience whilst trying to remain true to its original identity and mission. This point will be further addressed in Chapters 4 and 5, which look into the discursive nature of AJE's reporting both at the start of its operation and post-Arab Spring protests. Before that, however, there is a need to contextualise media power and media power contestation within the analytic framework of this study.

CHAPTER 3

DISCURSIVE MEDIA POWER

Given the complexity of media spheres today, as discussed in the opening chapter, alternative media do not only constitute a construction of 'otherness' *vis-à-vis* mainstream media, but have moved into the 'mainstream' in order to accommodate themselves with news media. Downing (2001) suggests, there is a blurring between 'mainstream' and 'alternative' media institutions, practices and discourses, which may suggest that differences are better understood as gradations on a scale of difference, rather than stark opposition. Nevertheless, unless a conceptual definition of what constitutes 'alternative' media is provided, such overlap risks blurring the lines between alternative and mainstream media to the point that the former term becomes meaningless. This problem has been taken up by Couldry, who has suggested that what is distinctive about alternative media is its ability to 'contest media power' (Couldry, 2003). In this book, I refer to media power as 'discursive media power', as I shall go on to discuss through the methodological approach employed in this study. Before that, however, the following section will unpack Couldry's framework for understanding media power as a conceptual resource that can be drawn upon to consider the extent to which Al Jazeera English (AJE) might be seen as providing 'alternative' discourses, as discussed in Chapter 2.

Contesting Media Power

According to Couldry, media power is defined as 'the concentration in media institutions of the symbolic power of "constructing reality"' (Couldry, 2000: 4). Couldry here refers to Thompson's (1995) categorisation of power where the media hold what is regarded as 'symbolic power', which is one of four other forms of power which he lists as 'Economic, political and coercive power'. According to Thompson, symbolic power can be found in cultural institutions such as 'the Church, schools and universities, [and] the media industries', and what these cultural institutions have in common is that they are a '[m]eans of information and communication' (Thompson, 1995: 17). Thompson refers to symbolic power as the 'capacity to intervene in the course of events, to influence the actions of others and indeed to create events, by means of the production and transmission of symbolic forms' (Thompson, 1995: 17). Couldry borrows the term 'symbolic power' 'because it emphasises that such domination always requires, as its medium, symbolic forms (speech, writing, performance, images, and so on), and its exercise may or may not involve explicit ideological content' (Couldry, 2003: 38). Additionally, 'it suggests there are power implications involved in such symbolic forms themselves, given that the ability to produce symbolic forms is not evenly shared' (Couldry, 2003: 38). While drawing upon Thompson's framework, Couldry nevertheless finds Thompson's definition of 'symbolic power' somewhat limited, on the grounds that its definition fails to include the 'wider impact that certain *concentrations* of symbolic power may have' (Couldry, 2003: 39). Instead, he suggests reconceptualising 'symbolic power' so it ties to other forms of power such as 'economic'. Thus, contesting Thompson's definition, he argues that:

> [a] *strong* concept of symbolic power, by contrast, would insist that some concentrations of symbolic power (for example, the concentration from which contemporary media institutions benefit) are so great that they dominate the whole social landscape; as a result, they seem so natural, that they are

misrecognised [. . .] and their underlying arbitrariness become difficult to see (Couldry, 2003: 39).

According to Couldry, it is this 'misrecognition' that renders the media's ability to construct 'social reality' so powerful. Media institutions hold the power to frame events by creating messages through symbolic forms. This is due to society's reliance on the media as a source for understanding the world around them. In his work with Curran, Couldry emphasises this point, arguing that 'media power remains a very significant dimension of contemporary reality. In short, media power is *an emergent form of social power* in complex societies whose basic infrastructure depends increasingly on the fast circulation of information and images' (Couldry and Curran, 2003: 4). This emphasis on media's ability to construct social reality is also a point of focus for Boyd-Barrett, who points out that the media 'portray, reflect, filter and negotiate the "real" world', such that the '"real" world, and [. . .] our ways of knowing that world, is influenced by the media and by widespread assumptions about the power of the media' (Boyd-Barrett, 2002: 16). This, he suggests, is achieved through the media's construction of the world through a process of 'selection, exclusion and inclusion'. This is a fundamental concept when understanding media power, since the process of 'selection, exclusion and inclusion' leads to media channels offering particular versions of reality.

Consequently, as a result of this 'selection, exclusion and inclusion', the media have the ability to *reduce* the boundaries of actual reality and set the limits accordingly. These limits become the parameters through which our understanding of the world takes place. Accordingly, a mediated space is created within which discourses circulate. Couldry describes this mediated space as a 'centralised' space, arguing that it is a basis of two myths, the first is the 'myth of centre' and the second is 'the myth of the mediated centre', a myth that is engendered by both academic and 'common sense' perspectives that view 'the media' as 'a highly centralised system of symbolic production whose "natural" role is to represent or frame that centre' (Couldry, 2003: 45). Couldry here argues that in reality there is no 'social centre' that can claim to represent all society's diverse values and beliefs, and therefore the media cannot be regarded as a representative and interpreter of that centre. He suggests that

> [t]he idea that society has a centre helps naturalise the idea that we have, or need, media that "represent" that centre; media's claims for themselves that they are society's "frame" help naturalise the idea, underlying countless media texts, that there is a social "centre" to *be* re-presented to us (Couldry, 2003: 46).

Accordingly, media power creates that centre and, as a result, a central platform or space emerges as a symbolic construct that has, nevertheless, real and significant social and political consequences.

Manuel Castells (2009) also articulates power through a dominant political sphere and that of 'micro-powers' which enter the sphere of domination, and as a result 'either the state changes or domination is reinstated by institutional means' (Castells, 2009: 15), which is an example of Fraser's model of 'subaltern counterpublics'. Castells elaborates on this in relation to discourse, where he states that '[a]lthough the emphasis here is on force, the logic of domination can also be embedded in discourses as alternative or complementary forms of exercising power' (Castells, 2009: 15), a point that will further be discussed in the later part of this chapter on Critical Discourse Analysis (CDA) as a tool through which power is exercised. Dominant discourses voiced within the mediated public sphere are those which have been selected by media institutions that hold the power. In other words, as Castells argues, '[w]hoever holds power [. . .] often decides what is valuable' since *value is, in fact, an expression of power*' (Castells, 2009: 28), thus creating 'social reality'.

The role the media therefore play in selecting and defining the news, in turn, defines social reality. Yet this social reality is also related to society's set of beliefs and values. On this, Couldry argues that dominant media practice not only enforces the creation of social reality, but also reflects the discourses which appeal to the majority of audiences. He defines 'media rituals' as 'any actions organised around key media-related categories and boundaries, whose performance reinforces, indeed helps legitimate, the underlying "value" expressed in the idea that the media is our access point to our social centre' (Couldry, 2003: 2). Thus, Couldry argues, media power is tied to the media's ability to capture the discourses which appeal to the majority of society, supporting a 'naturalisation' of media power. The construction of social reality, according to Couldry, is intrinsically a 'social process' where media power acts as a reproduction of the 'social practice' of its audience members. This is achieved by the media's 'circulation of images and discourse' that reproduce what is embedded in social practice. Accordingly what the media produce is not 'something superimposed on social practice from the outside; instead it is endlessly reproduced through the details of social practice itself' (Couldry, 2000: 5). As a result, the discourses which are reproduced by the media are a reflection of society's 'local patterns of belief' (Couldry, 2000: 5) where, by mirroring the discourses of their local social context, the media make these discourses 'natural' to their audience. As Couldry explains:

> These local patterns of belief and action have become so routine that, in practice, we run them together in a general conception of the media's 'effects' largely abstracted from those specific contexts of reproduction. As a result, the workings of the media's social authority tend to be hidden, and media power comes to seem natural (Couldry, 2000: 5).

In this way, media power constructs and re-constructs social reality, thus concealing its authority over society as a result of the naturalisation of the discourses it reproduces. According to Couldry, therefore, symbolic power is 'reproduced as legitimate' (Couldry, 2000: 3).

However, as Castells points out, the process of power operates through two forces, where one force seeks to 'enforce existing domination' and the other force resists dominant power 'on behalf of the interests, values, and projects that are excluded or under-represented' (Castells, 2009: 47). Hence, it is important to keep in mind that power is contextual, where these two forces 'ultimately configure the structure of power through their interaction. They are distinct, but they do, however, operate on the same logic' (Castells, 2009: 47). Hence from a methodological viewpoint, power and contestation of power can be potentially identified through a comparative discourse analysis which seeks to uncover discursive structures of power, and in turn highlight the forces of discursive media power.

Contestation of media power, according to Couldry, is a role taken on by alternative media, where they defy media power when 'media power is *not* reproduced as legitimate' (Couldry, 2000: 40). Couldry argues that alternative media have the capacity to contest 'the concentration of symbolic power in media institutions' (Couldry, 2002: 25). Thus, he argues that:

> in considering possible sites of resistance to media power, we must look not only at the distribution of economic and organizational resources and at contests over specific media representations of reality, but also at the sites from which alternative general frames for understanding social reality are offered. Beliefs in the media's central place in social life can be effectively challenged only by *alternative* frames (Couldry, 2003: 41).

According to Couldry, '*alternative* frames' are those that challenge the symbolic power of media institutions. Couldry and Curran use this concept to define alternative media as 'media production that challenges, at least implicitly, actual concentrations of media power, whatever form those concentrations may take in different locations' (Couldry and Curran, 2003: 7). While, as we have seen, Couldry remains insistent that symbolic power is not wholly distinct from, but instead connected to, other forms of power, this definition nevertheless suggests that, rather than to seek to define 'alternative media' in terms of their institutional form, we should look instead at where '*alternative* frames' occur.

At a global level, reflecting on Couldry's argument, it might be assumed that a central turning point of 'naturalisation' emerges as a consequence of the globalisation of media practice. Indeed, it may be the case that the process of naturalisation becomes a process of contestation as local traditions are made subject to distant forms of social, cultural and political influence. This possibility is suggested where Couldry highlights that, in relation to media power, 'there are moments and places where it is de-naturalised, perhaps even contested' (Couldry, 2000: 5). Contestation of media power, Couldry argues, occurs as a consequence of differing 'media cultures'. Thus, he suggests, 'if media power is as deeply embedded in social practice as I claim, then there will be significant differences in how that process works in different places, what we might call different "media cultures"' (Couldry, 2000: 5–6). Couldry and Curran have looked into 'how power is contested under different but structurally comparable conditions across the world' and argue that 'it becomes obvious that media power is rarely the *explicit* subject of social conflict', rather it is 'about relative control over society's representational resources' (Couldry and Curran, 2003: 4). In a mapping of various world conflicts, Couldry and Curran point out that '[w]hat was missing from those major conflicts was access by all sides to global means of self-representation, which could change the scale on which those conflicts were played out' (Couldry and Curran, 2003: 5). This highlights the importance of self-representation in global crises, where the construction of social reality in relation to such conflicts is affected by which 'voices' (Couldry, 2010) and 'interlocutors' (Fraser, 2007a) are represented within the mediated global public space, a point that will be explored in the analytic study in this book.

This study employs a CDA which is a suitable empirical approach for this research since it is an analytic method that investigates how discursive power is structured, reproduced and challenged. Discourse analysis is useful when studying relations of media power and possibilities of contestation. Therefore, this study uses this method to analyse AJE's news reporting and compare it to that of BBC World and CNNI and does so by adopting the 'thematic structure', as informed by Teun van Dijk (1985), as an instrument of analysis. The thematic structure has been adapted to suit the aims of this research, and has therefore incorporated a comparative analysis as part of this process, as I shall now go on to explain.

CDA as an Analytic Tool for Discursive Power

Many discourses through which individuals understand and engage with contemporary reality in modern societies are circulated through the media, which in turn act as a platform for the representation and discussion of particular issues. John E. Richardson (2007) defines journalistic discourse as follows:

> Journalistic discourse has some very specific textual characteristics, some very specific methods of text production and consumption, and is defined by a particular set of relationships between itself and other agencies of symbolic and material power. These three sets of characteristics − that is, the language of journalism, its production and consumption and the relations of journalism to social ideas and institutions − are clearly inter-related and sometimes difficult to disentangle (Richardson, 2007: 1).

Accordingly, a discourse analysis of journalistic texts allows us to better understand relations of power, as will be conducted in this study. However, before I go on to discuss the use of CDA as a tool for unpacking discursive power, I shall briefly provide a synopsis of the nature of this study.

This study employs a comparative approach between three transnational news channels as case studies, however, this is done through what I describe as a 'contextual' focus on AJE's performance. The reason behind selecting CNNI and BBC World as comparative case studies with AJE is due to the three news networks being 'tough competitors', where the 'BBC and CNN already have Arabic online services [and] [e]ach is keen to carve a niche in the others' market' (Barkho, 2007: 13). For this reason − along with the BBC and CNNI being news networks that AJE sought to emulate − these two dominant media channels become suitable case studies for a comparative analysis. Case study research aims to exemplify phenomena, and allows a grounded interpretation. Case study research is appropriate when there is a need to '(a) to define research topics broadly and not narrowly, (b) to cover contextual or complex multivariate conditions and not just isolated variables, and (c) to rely on multiple and not singular sources of evidence' (Yin, 2003: xi). For the purpose of this research, a case study approach is suitable since there is a need to gain a 'holistic' understanding of AJE's news reporting in relation to CNNI and BBC World. Such a comparative approach leads to identify nuances of media power relations and how they are possibly exercised by these three media organisations. However, there are limitations to case study research. One main problem with case study research is that it is not possible to generalise findings from one or two cases, since 'the case study contains a bias toward verification' (Flyvbjerg, 2006: 219), an issue that this research acknowledges. Overall, this study includes BBC World, CNNI, in addition to AJE as cases of globalised news coverage, and as a result, it is obviously not possible to claim that this is a 'representative' study. However it could be presumed that results are 'indicative'.

Data for this study was selected from two observational periods: the first period includes the initial year of AJE's operation, from 30 December 2006 until 30 December 2007; and the second period is post-Arab Spring from 30 December 2011 (a year after Mohamed Bouazizi's self-immolation) until 30 March 2014 (marking the end of the study period). The two observational periods were monitored in order to identify common news stories which would allow for a comparative analysis. Furthermore, the extended time periods were chosen in order to generate sufficient material for the analysis of AJE, CNNI and BBC's news, and to gain a better understanding of the discursive nature of these three networks. The news reports broadcast during news events were generally repeated throughout the day, so for each day of analysis there were mainly one or two reports for analysis for each news network. Two news stories were analysed in Chapter 4 as examples of AJE's news reporting in its first year of operation. The first news story analysed in this chapter is the kidnapping of the BBC Gaza correspondent, Alan Johnston, who was taken by a Palestinian group in return for the release of Muslim prisoners from foreign jails. The Johnston case was in the news from 12 March 2007 on the day Johnston was kidnapped, until 4 July 2007 which marked the day of Johnston's release. The second news event is that of the six Bulgarian medical workers who were sentenced to death in Libya for allegedly infecting over 400 Libyan children with the AIDS virus, which spans from February 1999 when the medics were arrested, up until the day of their release in July 2007. The analysis falls on the day the medics made an appeal during the court hearing on 20 June 2007, and the day the Supreme Libyan court announced upholding the death sentence on 11 July 2007. Chapter 4 looks into two news events that occurred in relation to the Arab Spring uprisings in both Egypt and Syria. The first of these news events is the sentencing of 528 members of the Muslim Brotherhood, on 24 March 2014 in Egypt's criminal court in Minya, for participating in an attack on a police station in mid-August 2013 in which a member of the police force was killed.

The second news event is that of the United Nations (UN) Security Council's veto on 19 July 2012 in New York, which failed to adopt a UN resolution on Syria due to the Russian Federation and China vetoing a British-sponsored resolution that would have imposed economic sanctions again the Syrian government for failing to carry out a peace plan. The reason for selecting the news stories, analysed in Chapters 4 and 5, was because they demonstrated examples of foreign involvement in a conflict situation in the Arab world, thus the news stories showcased local conflicts with a global interest. This made for excellent subject matter for the purpose of comparing AJE with CNNI and BBC World, since the news stories were of both Arab and Western interest. In both analytic chapters, analysis is not only concerned to identify how the stories were covered by each broadcaster, but in particular to highlight the degree to which AJE's discursive output served to reproduce, or provide an alternative, and possibly even contesting, perspective on the events in question. After explaining the parameters of this study, I shall now return to highlight the benefits and limitations of discourse analysis as a methodological tool for understanding and uncovering media power.

Discourse analysis, as a particular form of textual analysis that examines how discursive properties serve to construct textual meaning, has taken various forms and approaches throughout time. Historically, discourse analysis can be traced back to around 2000 years, from the time of Aristotle who explored the structure of discourse and its impact on the public. Discourse analysis developed further under the French Structuralists and Russian Formalists whose study of discourse was mainly connected with linguistic analysis. However, analysts began to argue that discourse analysis should include the study of the structure of the text as a whole (van Dijk, 1988: 18–22). The discourse analysis conducted in this study attempts to do just that where, in addition to the analysis of selected words and phrases, it also seeks to understand the overall meaning and direction of the text in the aim of uncovering any how examples media power is exercised. Philippa Smith and Allan Bell, therefore, define Discourse Analysis (DA) as:

> a close examination of text, including visual imagery and sound as well as spoken or written language. It is concerned with both the form of the text and its use in social context, its construction, distribution and reception. It aims to understand and elucidate the meanings and social significance of the text (Smith and Bell, 2007: 78).

The media have a role to play in the selection of discourses when producing a media text and thus also play a role in the construction of social reality, as discussed at the start of this chapter. So, slightly different to DA and '[m]ore specifically, CDA focuses on the ways discourse structures enact, confirm, legitimate, reproduce, or challenge relations of *power* and *dominance* in society' (van Dijk, 2003: 353). To deconstruct and better understand the term 'CDA', *analysis* refers to 'a detailed systematic examination of a particular object with a view to arriving at one or more underlying principles'; *discourse* is concerned with 'the way in which patterns of meaning

[. . .] are socially constructed'; whereas *critical* refers to 'the social effects of the meanings a reader is positioned or called upon to subscribe to in the act of reading, and the contestation of these meanings' (Locke, 2004: 9–10). CDA uncovers the underlying meanings that construct the various discourses that reflect 'the story of reality as it is presented to us through media or other cultural texts' (Van den Bulck, 2002: 85). Hilde Van den Bulck argues that '[t]here is no such thing as 'version-less' reality and discourse analysis tries to lay bare this constructedness. On the basis of this, a media or cultural text can be analysed by identifying the main discourses out of which it is structured' (Van den Bulck, 2002: 85). Since media power is defined in this study through its ability to construct social reality, CDA becomes an appropriate research tool for extracting discourses that assist in the formation of social reality and in turn highlighting relationships of power.

Accordingly through CDA, this study aims to identify the nature and qualities of discourse produced by each channel, in order to clearly assess whether AJE is offering a similar or different discourse to that of BBC World and CNNI. To the extent that AJE offers different discourses which challenge the discourses presented by BBC World and CNNI, then it might be seen as challenging the power of dominant global media, as suggested by Couldry's work on media power contestation. Studies on CDA have been primarily interested in notions of power, with more attention given to '*social power* of groups or institutions' (van Dijk, 2003: 354). Ruth Wodak and Michael Meyer explain this by stating: 'In texts, discursive differences are negotiated; they are governed by differences in power that is in part encoded in and determined by discourse and by genre. Therefore, texts are often sites of struggle in that they show traces of differing discourses and ideologies contending and struggling for dominance' (Wodak and Meyer, 2009: 10). Accordingly, since 'discourse is an opaque power object in modern societies', a strength of CDA is that it 'aims to make it more visible and transparent' (Blommaert and Bulcaen, 2000: 448).

Another particular strength of CDA that relates to this study is its ability as a method to extract and investigate discursive associations with ideology, since '[d]iscourse is seen as a means through which (and in which) ideologies are being reproduced' (Blommaert and Bulcaen, 2000: 450). Here ideology in relation to CDA is understood as 'patterns of language use and interlanguage/intervariety relationships that [carry] clear societal power or policy connotation' (Blommaert and Bulcaen, 2000: 456). The study of and analysis of language practice reflects on the ideologies that are being reproduced. This is of particular interest to this research, since the identification of differing forms of ideology which inform the practice of news networks helps unpack the power relation within the 'global public sphere'.

It is necessary to note, however, that there are limitations to a discourse analysis approach. Limitations to text-based analysis as highlighted by Greg Philo (2007) include its inability to show how competing discourses originate and relate to various social interests; how diverse social accounts compare to what is present and absent from a particular text; and what a text means to various audiences (Philo, 2007: 175). Moreover, additional problems in text-based analysis according to Philo lie in: '(1) the accuracy of representations, (2) the significance of texts to our own audience, and (3) the question of how rhetoric "belongs to" or is used by different social interests' (Philo, 2007: 175). For these reasons, Philo suggests the need to adopt a method that 'analyses processes of production, content, reception and circulation of social meaning simultaneously' (Philo, 2007: 175). While Philo's suggestion that discourse analysis alone cannot engage with such broader processes, it might be argued that attempts to study all aspects of communication simultaneously is not feasible, and that in this respect the delimited focus of discourse analysis might be seen as a strength rather than, or as well as, a limitation. Indeed, in response to Philo's critique, Elfriede Fürsich (2009) argues that text-based analysis is justified as a methodological approach in its own right since:

> media texts present a distinctive discursive moment between encoding and decoding that justifies special scholarly

engagement. The narrative character of media content, its potential as a site of ideological negotiation and its impact as mediated "reality" necessities interpretation in its own right (Fürsich, 2009: 238).

It should be acknowledged, however, that – like any methodological approach – discourse analysis does have limitations which should be both acknowledged and, where possible, addressed when designing a research methodology. For example, one particular criticism of CDA is that it does not factor the multiple readings of a text, nor does it take into account the various 'social circumstances' under which it is 'produced and consumed' (Blommaert and Bulcaen, 2000: 455). This raises the question of whether an analyst can speak on behalf of the 'average consumer of texts' (Blommaert and Bulcaen, 2000: 455), since analysts are seen to 'project their own political biases and prejudices onto their data and analyse them' (Blommaert and Bulcaen, 2000: 455). However, a strength of van Dijk's approach is that, through the application of a systematic and transparent method, such purely 'subjective' bias is minimised. Indeed, the use of van Dijk's thematic structure in this study as a methodological tool assists in adopting a systematic step-by-step approach rather than an 'interpretive' one through the process of *deletion, generalisation* and *(re-)construction*. This offers a transparent means of selecting and engaging with texts thus enabling a more apprehensible method for discourse analysis. In turn, this provides a frame that is systematically applied and from which a subsequent closer analysis of the discursive effect of particular combinations such as lexical choices, transitivity, modality, narrative combinations of words and images, and rhetorical tropes are developed (these terms are defined below). Accordingly, this allows for the emergence of a clear discourse to be identified where van Dijk's thematic structure bridges the gap between the microlevel and macrolevel of text, as I will go on to explain.

Other concerns regarding CDA as a methodological tool for textual analysis is the lack of consideration of time sequences of events, since it does not 'clearly explain the implications of previous discursive positions on subsequent ones', thus ignoring the 'historical nature of discourse' since '[t]exts always build on previous ones, taking up or challenging former discourses' (Carvalho, 2008: 163). Keeping this in mind, throughout the analysis in this study I highlight any historical and political contexts and connections to clarify points made within the news analysis. Another issue with CDA in relation to the study of news is that there have been concerns in the literature that there is the need to take into account the organisational and contextual forces affecting the textual output of news networks. In regards to this, Cottle warns that textual analysis on its own is not sufficient to 'understand why media representations assume the forms that they do as well as the silences found within media discourse' (Cottle, 2003: 5). Accordingly, there is a need to contextualise the findings of this research by considering the forces that influence discursive output. This is done in the final chapter where a discussion is conducted in light of the analysis outcomes through a consideration of how context is explicitly addressed in the interpretation of the empirical study, and through a consideration of factors influencing journalism in news analysis. Specifically, I consider some of the contributors of discursive reporting according to various factors that influence news output such as, among others, routines of *news sourcing*, the influence of the *culture of journalism* on reporting practices, and the nature of the *media organisation* itself and its relationship with its audiences. Accordingly, these elements – which I label in this book as *forces of discursive media power* – can be regarded as some of the contextual influences which need to be taken into account when making sense of discursive news output.

The 'primary topics' that have been addressed in the field of media through CDA can be identified as 'the narrative or sociolinguistic elements that construct or underlie news discourse', 'the implications of quotation and reported speech', 'the exercise of power, bias, and ideology in the press', 'the effects of the media in perpetuating social imbalance, notably racism and immigration', 'key genres, including broadcast interviews', 'the role of the audience', 'reception', and 'discourse comprehension and position within the media process', and 'issues of production and process of newsgathering and writing' (Cotter, 2001: 419). These topics reflect the various uses of CDA to news texts that are particularly useful in uncovering constructions of meaning, and in turn provides an appropriate methodological approach for uncovering nuances of power relations.

CDA, as a significant branch of discourse analysis, has become an area where scholars such as Norman Fairclough, Teun van Dijk, Ruth Wodak and Lilie Chouliaraki, have sought to develop and theorise through various frameworks. Norman Fairclough provides a three-dimensional framework for conceptualising CDA, where the first dimension is analysing 'discourse-as-text' (Blommaert and Bulcaen, 2000: 448). This form of analysis focuses on 'linguistic features and organization of concrete instances of discourse' through examples of 'choices and patterns in vocabulary', 'grammar' such as the use of passive and active verbs, 'cohesion' and 'text structure'. The second dimension is 'discourse-as-discursive-practice', that is 'discourse as something that is produced, circulated, distributed, consumed in society' (Blommaert and Bulcaen, 2000: 448). The third dimension is 'discourse-as-social-practice', which refers to 'the ideological effects and hegemonic processes in which discourse is a feature' (Blommaert and Bulcaen, 2000: 448). This is particularly useful for this research since is uncovers relations of power, as discussed above.

There have been various methods developed for discourse analysis that are specifically useful for the study of television news. Such frameworks employ a 'multi-modality' model of study, where 'the analytics of mediation applies a multi-modal discourse analysis onto media texts in order to study their visual and linguistic properties: camera/visual; graphic/pictoral or aural/linguistic' (Chouliaraki, 2006: 153). This form of analysis allows us to understand 'how the news text is put together in language and image' (Chouliaraki, 2006: 155). The multi-modal model provides a few aspects of analysis which need to be taken into consideration in television news analysis. One aspect is the 'mode of news presentation', which 'refers to the locations from which the news story is told and to the media that tell the story', and which includes whether the story is presented through the news anchor, or whether there is footage included in the news story through news reports, and whether it is archived or live (Chouliaraki, 2006: 160). Moreover, other considerations include whether studio presentation includes 'commentary of invited experts' or footage of 'oral testimony of witnesses from the scene of action' (Chouliaraki, 2006: 160). All these modes of presentation 'have an impact upon the ways through which the spectator comes to evaluate the news' (Chouliaraki, 2006: 160).

Other aspects of analysis which concern television news reporting include the 'verbal-visual correspondence' which needs to be taken into consideration when conducting television news analysis, and which has been employed in this study. The combination of the verbal and the visual lead to the creation of a social reality as constructed through the news report. This is confirmed by Chouliaraki, who argues that '[i]n order to be able to describe precisely how types of realism emerge through the multi-modal combinations of the news text, we need to talk more specifically about the work that language and image perform in the news text' (Chouliaraki, 2006: 161). This is necessary since the verbal 'tells us something about the degree of authenticity of a piece of news', since it is 'the verbal mode that establishes the distinct sense of reality that the story evokes for the spectator' (Chouliaraki, 2006: 162). Again, this notion of how social reality is constructed illustrates how media power is exercised within media texts, and through the CDA of television news 'linguistic and visual choices on screen [are] subtle indicators of the power of television to mediate the world to the world' (Chouliaraki, 2006: 166). For this reason, I relate to the usage of 'verbal-visual correspondence' within the analysis in this study, in order to gain a holistic understanding of the discourses intended by the three studied television news networks.

Whilst the multi-modality frame of analysis has been used in this study through the mode of presentation and the analysis of the voice against the image, this study mainly adopts van Dijk's 'thematic structure' which 'concentrates on the arrangement of themes in news reports, such as narrative patterns which creates dramatic tension in a story, or the non-chronological description of events influenced by its news value or relevance' (Smith and Bell, 2007: 86). Accordingly, the 'thematic structure' looks into bridging the gap between the microlevel and macrolevel of text. van Dijk makes a distinction between microlevel of analysis and the macrolevel of analysis and the role CDA plays in bridging the 'gap' between them:

> Language use, discourse, verbal interaction, and communication belong to the micro-level of the social order. Power, dominance, and inequality between social groups are typically terms that belong to a macrolevel of analysis. This means that CDA has to theoretically bridge the well-known 'gap' between micro and macro approaches, which is of course a distinction that is a sociological construct in its own right (van Dijk, 2003: 354).

Van Dijk's (1985) method of analysis through the 'thematic structure' is offered precisely as a means for bridging this gap, since it focuses on extracting the overall discourse of the text (macrolevel) and relating it to the selection of language and words (microlevel), which is what CDA as a method of analysis addresses (van Dijk, 1985; Fairclough, 2001; Wodak and Meyer, 2009). The reason behind this is to understand the discourse as a whole and the direction the text is taking, which in turn will assist in identifying whether there are discursive differences between AJE and the other two selected channels.

While van Dijk (1985) looks into the analysis of news in the press and not broadcast television news, a thematic structure analysis can be applied to television news since it focuses on the components of a news story that also make up the same components found in televised news broadcasts. The way in which the thematic structure is conducted is by summarising a detailed text into abstract and simpler meanings of text, which van Dijk explains can be done by what he calls the 'three summarising principles' which are 'deletion, generalization and (re-)construction', and which he labels as 'macrorules' (van Dijk, 1985: 76). The thematic structure is therefore conducted through the following steps:

- *Deletion:* Taking out any news items formed that can assist in reducing the text to an abstract theme, and which I refer to in this study as *omitting common news items.*

- *Generalisation:* Summing up the news items that are left into simple messages, which is a process I accordingly label as the *emergence of thematic narrative.*
- *(Re-)construction:* Organising the main news items of the story into a structured narrative which, in this study, is done through the analysis of the news reports and which is a step I label as *categories of analysis.*

I will explain these three steps further. After selecting the news reports, they were then transcribed with an explanation of the corresponding broadcast images alongside the text. This was conducted as part of Chouliaraki's (2006) 'verbal-visual correspondence', described above, which applies to television as a medium. This allowed for a multi-modal discourse analysis as part of understanding the visual and linguistic properties of the news reports. However in each chapter a full transcript of the report by AJE which included examples of counter-discourses was included in the text since such examples are strongly related to the focus of the research question. After the reports were transcribed, van Dijk's (1985) 'thematic structure' was applied in order to reduce the reports to an abstract form and in turn to assist in the comparative analysis.

'Deletion' – *Omitting Common News Items*

The first step in the thematic structure analysis is *deletion*. In this step, (a) common pieces of information – which I refer to as 'thematic discourse items' – across all reports analysed and discussed in the book were compiled. Thematic discourse items are the underlying structure of a news event, and therefore I use the term 'thematic discourse items' accordingly to refer to the micro-structure of the text. The classification of the news items was based on the categorisation of each news items as constituting a single piece of information. These thematic items were then (b) structured in a list, summarising the main stages of the news story from all the news data collected (see Table 3.1). Since the aim is to construct a comprehensive yet abstract

Table 3.1 Example of the 'deletion' process

THEMATIC DISCOURSE ITEMS	BBC World	CNNI	AJE
DAY 1			
Account of Johnston's kidnap	X	X	
Hamas pressuring to secure release	X		X
DAY 2			
Hamas quote: Zahar	X	X	
Hamas quote: Haniya			X

The 'x' marks the channels which included the corresponding thematic discourse item.

The shaded rows highlight the thematic discourse items which were broadcast by all three channels.

list in order to identify and eliminate the similarities, different pieces of information which hold the same meaning were grouped under one news item. So, for example, in the Johnston case, the BBC World report said that Johnston was 'taken at gunpoint', whereas the CNNI report said that he was 'taken on his way to work'. To avoid a long list of items with many overlaps, both statements were combined under the heading of 'Account of Johnston's kidnap'. In another example, one AJE report said that 'Hamas issued an ultimatum' to the kidnappers, whereas the BBC World report said that Hamas exerted 'pressure on the kidnappers', so again both statements were combined under 'Hamas pressuring to secure release'. Additionally, there were cases where despite combining statements under one heading, there was still a need to make a distinction. So in the Johnston case, BBC World and CNNI included a quote from one Hamas official and AJE included a quote from another Hamas official, and since both quotes were along the same lines, they were grouped under 'Hamas quote'. Yet in this particular example, a distinction is still made in the table, since there could potentially be a reason behind each channel using different people for quotes (see Table 3.1 above). From this systematic grouping of information, as the shaded area in the table shows, a comprehensive list of news items for each news story was generated. The table above exemplifies how the items were tabulated.

After compiling the news items of each news narrative, the thematic items that were featured by all three channels were highlighted in grey. The reason behind this was to eliminate elements of the news narrative that were reported by all three channels (highlighted in grey), thus allowing for the differences in reporting to emerge in preparation for the second step of the thematic structure, which is the *emergence of the thematic narrative*. The remaining news items were, therefore, included in the news reports based on an editorial decision made by each channel (whether conscious or unconscious), which in turn reveals the selection of information adopted by each channel. Therefore by highlighting the common items and deleting them, according to van Dijk's *Deletion* approach – which means extracting any items that can assist in reducing the text to an abstract structure – the discrepancies in reporting become more apparent. Thus the deletion of the common news items made way for the emerging discourse patterns to become more visible.

'Generalisation' – *Emergence of Thematic Narrative*

In order to understand and visualise the emerging discourses that each channel focused on in its news reporting, the remaining thematic news items resulting from the previous step of *deletion* were assembled into another table in line with van Dijk's principle of *generalisation*, which structures the remaining news items into abstract thematic narratives. Each column provides a narrative from the remaining information as selected by each channel. Reading down each column (see Table 3.2) produces an understanding of what each channel was reporting on, thus making it clearer which discourse each channel adopted.

As noted from Table 3.2, this tabulation leads to the emergence of vertical patterns in the form of clusters. By reading the clusters formed in every column, a discourse by each channel is detected. So for example in the Johnston case, CNNI focused on showing tapes sent by the kidnappers and on recordings of Johnston wearing an explosive vest. Such examples indicate that CNNI is framing this event through a discourse of 'terrorism'. Through this process it

Table 3.2   Example of the 'generalisation' process

BBC World DAY 1	CNNI DAY 1	AJE DAY 1
Palestinian journalists' rally		Palestinian journalists' rally
	Longest-held hostage	Longest-held hostage
		Archival images of Johnston reporting
Hamas pressuring to secure release		Hamas pressuring to secure release
Kidnapped 12 March	Kidnapped 12 March	

became apparent which discourse each channel was presenting, thus making the distinctions in reports clearer. Such an exercise made it easier, for the purpose of this study, to reach an abstraction of the report which revealed a distinct discourse, which was then used as a basis for a detailed analysis during the process of *(re-)construction*. The process involves analysing the words, sentences, and overall meaning of the text though the lens of the discourse which emerged from the previous steps of *deletion* and *generalisation*.

In addition to a vertical reading of the emerging discourses from Table 3.2 above, the table also provides a visual comparison between the news items each channel reported on. By reading horizontally, a comparison between which channels included the same news items can be made. So for example, in the above table, the first component in the CNNI column 'Longest-held hostage' is kept on the same line as the 'Longest-held hostage' component in the AJE column. Thus, by reading across the table, a comparative understanding of which piece of information was featured by which channel also becomes apparent.

Overall, this second step of the thematic structure is crucial, since the identification of a particular discourse by each news organisation means that each channel is adopting a specific position on the news story studied. What is particularly interesting to note as part of this step, is that a clear thematic narrative, which consisted of a clear identifiable discourse, emerged by each channel on both stories analysed. In turn, the emergence of distinct discourses through this process of *generalisation* makes way for the next step of analysing and comparing the discourses that emerge from the reports in further detail through the process of *(re-)construction*. Through this step, an analysis of words and images that were included in the original reports are conducted through a comparative analysis, keeping in mind the discourses which emerged from the process of *generalisation*. It is this step of bringing details and context back in the analysis that prevents the method becoming a mere abstraction.

'(Re-)construction' – *Categories of Analysis*

As mentioned earlier, this study is concerned with bridging the gap between the macro/micro-elements of the text in order to reach a holistic understanding of the emerging discourse by each studied news network. Therefore a close account of the different elements that make up a text is essential in the construction of the overall meaning of the text. To identify these elements for the purpose of the analytic study, Jenny Kitzinger's (2007) classification of key 'cues' found in news reporting is useful, and these include:

- Images used
- Type of language used
- Labels and definitions employed
- Explanations offered
- Responsibility assigned
- Solutions proposed
- Narrative structure
- Contextualization and links
- Historical associations invoked
- Similes and metaphors
- Emotional appeals
- Who is invited to comment
- How different speakers are introduced
- How different characters, groups, social movements or entities are described (Kitzinger, 2007: 141–142).

To these categories, I add a category which includes ways on How the report is introduced, where during the process of study it was found that the news anchor often 'frames' the overall presentation of the report, and in turn the meaning that is produced. The 'cues' listed above are used as categories of news analysis within the final step of analysis. These categories help to uncover how discourses are structured by each news channel. Under these categories for analysis, which I use in this study, reports on each news story are reconstructed and compared through the lens of the discourses that have been identified for each of the studied channels. In this way the analysis incorporates all details, be it verbal or visual. Accordingly, these categories for the analysis of news reports have been used as main categories under which the analysis of the data has been conducted.

In addition to the above cues, van Dijk, also offers several points that are used as a basis for this analytic and comparative study as sub-categories of analysis (van Dijk, 1998: 31–35):

- *Lexical items* – Lexical items are words that 'generally or contextually express values or norms' (van Dijk, 1998: 31), as part of a lexical pattern. The study of words in the comparative analysis includes identifying which words each channel used to describe an event or group. Accordingly, this identification of words projects the meanings and messages intended by each channel. At the same time, the study of words by themselves is not sufficient, and it is therefore necessary to look into the context as the following areas suggest.
- *Propositions* – The study of propositions involves meanings that are derived through structures of sentences. It is necessary to look into lexical items within the sentences they are situated in, and therefore understand the context the word is used in.
- *Implications* – Here the meaning is in fact implied rather than explicitly stated within the text. Therefore, in this case, the meaning would need to be extracted from the text.
- *Descriptions* – This takes the analysis to another step where the events could be described generically or specifically. So in this case

the analysis focuses on the 'sequences' of propositions in the text, thus looking into the overall text.

The above concepts have been deployed in the third step of the thematic structure of which is the *(re-)construction* step, where the identified discourses which emerged from the first two steps of the thematic structure were further analysed through lexical items, propositions, implications, and descriptions. Such analysis relied on the discourse as a context for decoding the above elements of the text, and in turn asserting and emphasising the identified discourse for each channel. Other tools of linguistic analysis that have been deployed in this study include *transitivity, modality, presupposition, rhetorical tropes*, and *narrative* (Richardson, 2007). These tools are defined as follows:

- *Transitivity* is an analytic tool which 'describes the relationships between participants and the roles they play in the processes described in reporting' (Richardson, 2007: 54). Transitivity relates to representation where a process includes three components: the *participants* who are 'involved in the process', the *process* itself which is articulated in the verb, and the *circumstances* which such as the location and time (Richardson, 2007: 54; Haig, 2001).
- *Modality* relates to 'judgements, comment and attitude in text and talk, and specifically the degree to which a speaker or writer is committed to the claim he or she is making' (Richardson, 2007). Here the speaker or writer is 'attempting to persuade us of their representation of reality' (Haig, 2001). Therefore, through modality, further steps can be made in the analysis which assist in revealing links between content and function (Richardson, 2007: 59).
- *Presupposition* can be regarded as a 'taken-for-granted, implicit claim embedded within the explicit meaning of a text or utterance' (Richardson, 2007: 63). Here the meaning is related in a way that assumes that the information is known or 'presupposed'. According to van Dijk, '[t]hey may be strategically used to obliquely introduce

into a text propositions which may not be true at all. This is also the case for presuppositions that embody opinions' (van Dijk, 1998: 34). An example of this is using the word 'illegal' for asylum seekers. There are three 'linguistic structures' associated to 'presupposed meaning': the first is through certain words such as 'change of state verbs'; the second is through the definite article (the) or the possessive articles (his or her); and the third is through 'wh- questions' such as why, who and where (Richardson, 2007: 63).

- *Rhetorical Tropes* refer to words which are used 'to denote-connote something apart from their ordinary meaning' (Richardson, 2007: 65). This is considered a play on words, it serves to provide a linguistic imagery that the speaker or writer would want to convey.

- *Narrative* is regarded as 'the contents of news stories and the ways that such news stories are presented' (Richardson, 2007: 71). The structure that a narrative usually takes in hard news is through a 'three-part plot structure' as follows: '*complication* (that actual reported "event"), the *setting* or background and then the *outcome* to the story' (Richardson, 2007: 71). In the analysis of television news, however, it is hard to conceptualise narrative as a story-telling process rather 'it is arguably more revealing to identify some pervasive characteristics of the texts of television news and then to elaborate from these characteristics some core principles of their discursive intelligibility' (Montgomery, 2007: 91), and it is through this conceptualisation that I refer to 'narrative' in this analytic study.

The above linguistic tools are applied in the analysis and examples of these tools will be outlined during the analysis. In addition to these tools, I also include *nominalisation*, which 'is a transformation which reduces a whole clause to its nucleus, the verb, and turns that into a noun' (Fowler et al., 1979: 39). In CDA, nominalisation 'is a way of representing a process as a noun, as if it were an entity, which has the effect of obfuscating agency and causality, and the time at which processes take place' (Haig, 2001: 215). Another tool that is applied to the analysis is *overtextualisation*, which refers to the 'variety of terms to describe the same action' and 'often indicates areas of intense preoccupation in a text, and hence in the ideology of the writer'

(Haig, 2001: 214). These linguistic tools provide the study with a detailed and holistic understanding of the discursive message and assists in the identification of how media power is exercised in news texts. In relation to the overall analysis, I also refer to the concept of framing during my analysis, which connects to 'the process of culling a few elements of perceived reality and assembling a narrative that highlights connections among them to promote a particular interpretation' (Entman, 2009: 336). Framing, therefore, relates to discourse analysis as a methodological process in that it aims to provide a particular 'interpretation' to the text.

In addition to the areas of text and images outlined above, there is also a need to take into account the delivery of the reports which includes the tones used. According to Nuria Lorenzo-Dus (2009), the way in which a journalist delivers the report has a significant impact on the way it reaches the audience, and in turn has an impact on the message. Lorenzo-Dus clarifies this further:

> The purchase of emotional expressiveness on television is not only a matter of frequent instances of verbal sharing on our screens [. . .] it is a matter of presenters' direct look and address to camera, of upbeat/downbeat tunes that punctuate the happy/unhappy stages of stories told in documentaries and of a combination of concrete and metaphorical silence in live news to garner a particular mood. It is, all in all, a matter of television discourse drawing upon a wide range of semiotic resources in order to foreground emotion – to create the impression that it talks to us, with us and about us in ways which we can not only hear and see but also, and crucially, feel (Lorenzo-Dus, 2009: 184).

What this entails is the consideration of emotional appeal that has become standardised in many reporting styles, and considerably so in Arab news reporting. The discourse analysis conducted in this study will take this into account. Finally, at the end of the three steps of the thematic structure conducted in each of the analysis chapters, the comparative results are then discussed and summarised.

The above has outlined the steps of the thematic structure that will be applied to the empirical study. A comparative study leads to understanding how each station reported on the two stories selected for analysis, thereby allowing for differences and similarities to be revealed. Consequently, this will assist in positioning AJE within the 'global public sphere' in relation to BBC World and CNNI. The difficulties of such a study, as discussed earlier, relate to the extent of 'objectivity' that can be achieved by the researcher when reading the discourses projected by each channel. This is why analysis through the thematic structure is a useful research process, as it allows for a systematic step-by-step uncovering of the discourses through each informational component. Hence, rather than 'reading off' texts through the subjective impressions of the researcher, this offers a more transparent and reproducible framework of analysis.

CHAPTER 4

BEFORE THE ARAB SPRING

As discussed at the start of this book, the proliferation of satellite television news over the last two decades has stimulated substantial debate over the degree to which the field of transnational global news serves to extend hegemonic perspectives and power, or enable an increasing number of voices and perspectives to gain representation. This chapter explores how satellite news is constructed through power relations between various transnational news networks through a discursive and visual analysis of their news reporting. Such a study of news content allows us to better understand the role these news networks play as political institutions in the mediatised global sphere. This chapter explores whether Al Jazeera English (AJE), as a relatively recent entrant to the field of international news, provides additional and/or alternative mediated discourses on regional issues to a global audience and, accordingly, plays a role in expanding the 'global public sphere'. In doing so, it seeks to critically consider claims made by AJE itself that it serves to present alternative perspectives on international news; as well as positions in academic debates that have suggested that an increase in the number of satellite television broadcasters from the developing world is facilitating an increasingly democratised field of global news. The following analysis will now apply the steps of study as outlined in Chapter 3.

The Kidnapping of a Journalist

The journalistic construction of the Alan Johnston case is 'newsworthy' due to the involvement of the media itself in this political story. The kidnappers took a member of the international press corps to attract international attention. It is also worth noting that at the same time, there were other hostages who were taken captive in the Middle East, but this news story was one that gained an increased attention due to the captive being a BBC journalist. While this news story includes a sequence of events, the period of analysed news broadcasts for this study focuses on two of the last events in the story, namely the 100th day vigil organised by the BBC and the day of Johnston's release. Although the day of Johnston's release attracted the bulk of the media coverage, the two days of reporting shed different light on the Johnston case: the 100-day vigil reporting takes a sombre tone, and the day of release is more celebratory.

The story of Alan Johnston's kidnapping may be summarised as follows: Johnston had been working as a BBC reporter in the Gaza Strip for three years until the time of his kidnapping. He was the only remaining Western reporter residing in Gaza, which meant that if a Western journalist was to become a target it was he. Johnston was kidnapped on his way home from the office in his media car on the streets of Gaza by a Palestinian militant group called Jeish Al Islam, translating from Arabic to 'The Army of Islam'. During the period of his kidnapping, which lasted 114 days (around four months), Johnston appeared on the group's website twice. The first tape showed Johnston wearing red and asking for the release of Muslim prisoners from foreign jails, and the second time he was wearing an explosive jacket. During the period of kidnapping the British Consulate in Jerusalem was negotiating his release with Hamas, which was the newly elected Palestinian leadership at the time, and other political fronts; and it was Hamas who was able to secure his release after reaching a deal with the kidnapping group.

The three stages of the thematic structure have correspondingly revealed the following three points in relation to the Johnston case. In the first step of the thematic structure, the deletion process involved

Table 4.1   Common thematic items of the Johnston story used by all three channels

Common Thematic Discourse Items
100 days in captivity
Worldwide vigils
Name of kidnapping group
End of 114 days/4 months of captivity
Release happened in early morning
Johnston quote
Hamas quote
BBC Jerusalem bureau celebrations
Johnston meeting British Consul General

the reduction of the news text by eliminating the thematic items of the news story that were similar across all three channels. These items included in the reports provided, more or less, the same basic news items across the three channels (see Table 4.1 above). This reflects that global news networks provide similar factual information, however the discursive contexts vary, as we shall go on to see.

This leads to the second point where although the 'factual' information is similar, the construction of the story varies from one channel to another. Thus the selection of which secondary information is to be included and omitted reveals the editorial decisions made during the making of the news reports. It was noted that this editorial decision on what angle to pursue in the structure of the news reports reflects each channel's political-economy be it a British, American or Arab representation. For example, the BBC, as a British broadcaster and also an actor in the story, focused on the diplomatic relations between Britain and the Palestinian leadership, insinuating that the kidnapping was not an official Palestinian position. This was evident in the use of government sources where reports adopted an editorial stance consistent with the UK government. On the other hand, CNNI strongly focused on highlighting the terrorism discourse by making connections to the kidnappers and Al Qaeda, which may also be 'indexed' to government agendas or appealing to popular concerns regarding terrorism. As for AJE, it was keen to emphasise that the diplomatic relations between Britain and the Palestinians were not tarnished, in addition to highlighting the good will of the Palestinian leadership and the efforts it went through to secure Johnston's release. So although there is an area where all three channels overlap in their reporting, there is a distinctive version of the story that each channel brings to the global audience and which was revealed through the thematic structure process.

On the 100th day of Johnston's captivity, the BBC report includes most of the elements that are of interest to them. Interestingly, the report highlights that there is a difference between the Palestinian kidnapping group Jeish Al Islam and that of other Palestinians by showing the images of rallies that were conducted by the Palestinian journalists who are heard chanting 'Free Free Alan'. Here the reporter points out that this rally was held '[i]n the days immediately after' the kidnap when the Palestinian journalists 'took a leading role demanding his release'. Here the lexical choices 'in the days immediately after', refer to the urgency of the action through both the word 'days' enforced with the word 'immediately', thus emphasising the keenness of the Palestinian journalists to support Johnston's release. To add, it is notable here that by describing the Palestinian journalists through the lexical choices of taking a 'leading role', it portrays these Palestinian journalists as taking initiative in this case, thus showing their support for Johnston and their disapproval of the kidnapping, and in turn distancing them from the Palestinian kidnapping group. It is worth pointing out that the news anchor introduces the kidnappers in this story as 'a group calling itself "The Army of Islam"' without any descriptions such as 'terrorist group' thus giving them a politically neutral labelling through the use of the lexical choice 'group'. This portrayal reflects that the BBC's discourse is distinguishing between the kidnapers and the Palestinian people without also criminalising the kidnappers, which serves to focus on maintaining positive diplomatic ties with the Palestinian front as we shall go on to see. Accordingly, the BBC appears to be advocating the goodwill of the Palestinian people, which is interesting to point out since the BBC is a victim in the specific news story.

Figure 4.1   Image from the BBC of masked kidnappers (BBC World, 20 June 2007)

The inclusion of the kidnappers' website, in the BBC report, with the kidnappers appearing wearing black hoods (see Image 4.1 above) brings the viewer's attention to the dangers that Johnston is going through. Images of the kidnappers making online demands are typical of a terrorism scene, and thus serve to identify this group as terrorists. The kidnappers are shown here requesting the release of Muslim prisoners from foreign jails. Alongside these images, the reporter points out that 'the group also stressed that negotiations are going on'. The use of the lexical item 'stressed' in this proposition serves to emphasise that the group is keen to negotiate, thus bringing some hope to this part of the report despite the images showing a more sombre representation. This 'verbal-visual correspondence' (Chouliaraki, 2006) provides both the sense of danger through the images yet softens this affect through the proposition that the kidnapping group is stressing that a solution is underway. It is interesting to note that in this example, although the selected images provide a factual representation, the commentary which accompanies these images helps frame the report towards a different specified discourse. So although, it is interesting to note the BBC in this report includes images of the kidnappers making online demands,

Figure 4.2   Image from AJE of Palestinian journalists' rally (AJE, 20 June 2007)

where the inclusion of the kidnappers' image reminds the viewer of the terrorism frame, it points out that a diplomatic solution is underway. This again emphasises positive diplomatic relations, which is a point that the BBC stresses throughout its reporting on the Johnston case.

The AJE report that day also shows images of the Palestinian journalists' rally holding plaques with 'Free Alan' on them (see Image 4.2). Interestingly CNNI did not include this shot, although the BBC did. However, like the BBC, AJE as an Arab broadcaster seems keen to portray Palestinians as disapproving of the kidnapping and calling for the release of Johnston. Accordingly, the report ends by mentioning that Hamas is playing a role in trying to secure Johnston's release, issuing an ultimatum to the kidnappers two days previously: 'Hamas has issued an ultimatum demanding his release two days ago but nothing's been heard from the group calling itself "The Army of Islam" since then'. This information is not mentioned in the BBC World or CNNI reports but is an important recent act by the Palestinian leadership and one that shows the active role Hamas is playing in this story, and which is also a significant local element of the story. AJE is, therefore, emphasising that the Palestinian journalists and Hamas do not support the kidnapping of Johnston, despite the kidnappers being Palestinian. This is an important point for AJE to make which, similar to the BBC, is keen to distance the Palestinian journalists and Hamas from the kidnapping group.

It is also notable that the AJE report ends with images of Johnston making an online appeal in red, similar to the CNNI report. At first sight it is questionable whether AJE is framing the story through a terrorism discourse, since this image is broadcast through AJE. Nonetheless, it becomes apparent that this is not the case throughout the day of Johnston's release, as CNNI expands on the terrorism discourse and AJE focuses on the role of Hamas in securing the release. Arguably, also, this image is not as confronting as the image displayed by the BBC which shows the images of the kidnappers covered and holding guns. This BBC image is one that global audiences have associated with images of kidnappings and beheadings conducted in Iraq.

Overall, there seems to be some overlap between AJE's coverage and that of the BBC and CNNI through the inclusion of similar footage. Yet the overall framing of AJE's reporting through differing discourses, offers a different reading to these images. The AJE report presents a discourse similar to that of the BBC, which is distancing the Palestinian leadership and journalists from the kidnappers. Each news network, however, is doing it from differing positions: AJE being an Arab broadcaster is keen to focus on how the Palestinians are disapproving of the kidnapping act, and the BBC being on the other side of the news story is also keen to promote this point as an emphasis of positive relations between the Palestinian and British positions. Interestingly, it can be argued that both the BBC and AJE are practicing public diplomacy yet for different reasons. This is not to suggest that AJE is being dictated by any government position, but rather that AJE – as an Arab broadcaster which has historically supported the Palestinian cause – is keen to show that the act of kidnapping foreign correspondents is not one which is approved by all Palestinians, thus distancing the Palestinians from the terrorism discourse.

On 4 July 2007, 14 days after marking Johnston's 100 days in captivity on 20 June 2007, Johnston's release is announced and unlike the reports marking the above event this coverage is extensive and breaking news. Since this is a major event, all three channels feature long reports and continuous coverage. In these reports, all three channels feature similar footage of Johnston being escorted to his car after his release, with chaotic scenes of media around him trying to catch a glimpse of him in order to get a quote. In addition to this scene, a press conference is held on the day by Johnston and the Hamas leadership. Furthermore, some reports feature Johnston having breakfast with Hamas senior officials, a sign of positive diplomatic relations.

The BBC's report on the day of Johnston's release is, not surprisingly, delivered with a celebratory tone in response to the news of the safe release of a BBC employee. A quote by Johnston sheds further light on what happened to him during his confinement where Johnston talks about the uncertainty he felt during his captivity. Here he uses propositions that are quoted frequently in other coverage broadcast that day which describe his feelings, such as being 'removed from the world' and 'like being buried alive'. Here the rhetorical tropes 'removed' and 'buried', by implication, serve to emphasise how frightening Johnston's experience was. Other propositions that are stated by Johnston include: 'in the hands of people who were dangerous and unpredictable', emphasising that the kidnappers were a frightening group. Although the report states that Johnston had not been beaten, Johnston describes incidents of violence that he endured, and provides details:

> About three in the morning, on the first night, they rolled me up and put a hood over my head, and handcuffed me and took me outside, and of course you wonder that way how else it's going to end.

This detailed description and multiple quotes from Johnston on his agonising experience during his captivity highlight the terror that Johnston endured from the kidnappers, thus stressing the risks and dangers involved in this kidnapping. This point further builds towards the importance of this story, and focuses on the dangers associated with kidnappings. This is the first account we hear from Johnston since his kidnapping, so it is an important aspect of the report to include especially since Johnston is a BBC reporter.

The next part of the BBC World report focuses on the diplomatic relations between Britain and the Palestinian leadership, and includes a positive framing of Hamas. The report states that Hamas secured Johnston's release and includes a quote from senior Hamas leader, Mahmoud Zahar, who stresses that Hamas would not allow such actions to happen in the future: 'It's a clear message: we will not allow illegal actions against anybody, we are going to make law'. This quote, where the Hamas leader is identifying the kidnapping as an 'illegal' action, shows how Hamas is disapproving of the kidnapping. This lexical choice of 'illegal' by Hamas to describe the kidnappers' actions aims at distancing Hamas from the kidnappers. The report tells how it was Hamas who helped release Johnston despite its negative image in the Western world: 'But it was this organisation, shunned by the international community, which did most to secure his release'. In the first clause, the transitivity participants here are 'organisation' and 'international community', where the action verb is 'shunned' which reflects a strong notion of rejection on behalf of the international community towards Hamas thus emphasising how the world disapproved of Hamas at a global level. This comment is made against images of Johnston having breakfast with the Hamas leadership, where this verbal-visual correspondence showcases the good relations between Johnston and the Palestinian leadership, and almost frames Hamas as legitimate as opposed to being 'shunned'. This comment places Hamas in a positive light, and the fact this statement comes from the BBC, a party directly affected, makes it all the more notable. Here, the manner in which the BBC presents Hamas in a positive light further feeds into its discourse of the positive relations with the Palestinian leadership. This is emphasised even further by a statement taken from the British Foreign Secretary, David Miliband, naming the Hamas leaders and recognising 'the priority that's been given to the issue' by Hamas. This recognition of

Hamas's efforts by a top British official, aligning with the discourse of British/Palestinian relations adopted by the BBC, suggests that the latter's discourse is itself supported by the British government's diplomatic stance.

The report that was broadcast on CNNI that day incorporates more extensive coverage than that of the first event. Before the report is presented, the news anchor provides a general recap to the story as an introduction to Johnston's release. However, in these few lines there is a mention of Johnston wearing an explosive vest where the anchor says: 'Very little had been heard about him, apart from two tapes, the second of those tapes showing him enwrapped by a suicide – what looked like a suicide vest'. The repetition of the lexical item 'suicide' in the second clause acts as a confirmation that Johnston is in danger. Additionally, the interruption in the flow of the proposition tends to place additional emphasis on this image of the suicide vest through the use of the lexical items 'what looked like', thus linking the commentary to the image of Johnston in the suicide vest through a verbal-visual correspondence. The modality presented in this example, where there is a focus on the word 'suicide' by repetition, serves to present the story through the terrorism discourse, where the anchor is not just conveying information but presenting it through a terrorism framing. It is interesting to note, therefore, that although the occasion of this event is Johnston's release, the report is introduced with a focus on the suicide vest which is not part of the events of that day altogether. This form of presentation through the verbal-visual correspondence used where the inclusion of Johnston wearing an explosive vest and stressing the word 'suicide', makes the story newsworthy according to CNNI since this places the Johnston incident within the broader frame of the war on terror. Here CNNI from the start is advocating the terrorism discourse, and this is evident throughout its reporting on this story.

The CNNI report starts with the words 'Free at last!', which is a description of Johnston's current situation, and then follows by stating that during the length of the period of captivity Johnston endured: 'one hundred and fifteen days in captivity – one hundred and fifteen days of fear'. Although the first clause of the proposition is a simple statement of fact since it states the exact number of days

Johnston was in captivity, the second clause of the proposition restates the number of days by substituting the lexical items 'in captivity' with the lexical choice of 'fear' thus providing a description of Johnston's experience. This presupposition that Johnston's experience was terrifying is carried out through the substitution of the words 'in captivity' with 'fear' thus marking a move from a statement of fact to a suggestive claim. A statement of quality, therefore, is assigned here through the word 'fear' connecting the captivity with terror, thus reiterating the terrorism frame adopted in this story. A quote selected from Johnston serves to support this where he narrates how there was talk of him being killed by his captors during his captivity: 'they chained me up for the twenty-four-hour period, they talked about the possibility of killing me over the next two days'. The possibility of him being killed is raised for the first time here, as it is not mentioned in the analysed BBC World reports studied that day. This serves to reaffirm the previous proposition of 'one hundred and fifteen days of fear' as a credible claim. This, therefore, connects the group to terrorist violence as it implies that the group did not just hold him for negotiation purposes. Although it would have been expected that the BBC would include this quote as an affected party, CNNI has opted to include it as a selection that feeds into its choice of the terrorism discourse.

The CNNI report then provides some background to the kidnapping and introduces the group by describing them as 'shadowy' through the proposition 'Johnston was kidnapped in Gaza on the 12th March by a shadowy group called "Jeish Al Islam" − The Army of Islam'. The mode of this proposition is mainly unmarked apart from the lexical description of the group as 'shadowy' which through modality places the kidnappers in a negative frame, thus further emphasising and advocating the terrorism discourse. This choice of words becomes notable when compared to AJE's use of a more politically neutral description of the kidnappers in the analysis of the AJE report to follow.

It is worth noting that in this report, CNNI has chosen to include all three tapes released during Johnston's period of kidnapping: the tape of Johnston dressed in red saying that he is being treated well and is in good health; the second tape where the kidnappers are shown covering their faces and holding machine guns – similar to the tape shown in the BBC World report; and finally the tape where Johnston is making a plea in an explosive vest. It is notable that, alongside the image of the masked kidnappers, the reporter points out that the kidnapping group is requesting that Britain release Muslim prisoners from foreign jails, a point also mentioned in the first BBC World report. However, the CNNI report goes further, outlining the kidnapper's request for the freeing of 'Al Qaeda's spiritual ambassador in Europe'. This mention of Al Qaeda, which is the group linked to the September 11 attacks against the US, associates the kidnappers to this top terrorist group. This contextualises the kidnappers as part of the wave of terrorism that the Western world is fighting, and by framing them in this way the terrorism discourse is further emphasised.

The third tape used in the CNNI report is of Johnston in an explosive vest (see Image 4.3), and is accompanied by the use of the lexical item 'chilling' to describe the tape: 'But the most chilling video came a few weeks later when he appeared to be strapped with a bomb'. This proposition uses the lexical choices 'most' and 'chilling' as a form of overtextualisation where the lexical item 'most' provides a superlative to the lexical choice 'chilling' to accentuate and intensify the effect of fear. This is then followed by a snippet of the video where Johnston is quoted as saying: 'As you can see I've been dressed in what is an explosive bomb which the kidnappers say will be detonated if there's any attempt to storm the area'. This proposition carries a threat to the viewer, however since this is being included in a report broadcast on the day of Johnston's release this threat no longer carries a sense of present danger. It is therefore questionable why there was the need to include Johnston's quote that day, which seems to be more of an assertion of the terrorism discourse which CNNI is adopting to frame this story.

The few examples of verbal-visual correspondence in the above analysis emphasise the terrorism discourse, such as the image of Johnston strapped to a bomb and the reporter describing this as 'chilling', in addition to the images of the kidnappers holding weapons and the reporter linking their requests to the release of Al Qaeda's spiritual leader. This selection of words and associations against the

Figure 4.3   Image from CNNI of Alan Johnston in explosive vest (CNNI, 4 July 2007)

images shown, frame these images through a terrorism frame. In comparison to BBC World and AJE, which also show images of Johnston and his kidnappers, there is more of an emphasis by CNNI to present Johnston's kidnapping through a popular American discourse of terrorism. Although the connection between Johnston's kidnapping as an act of terror is legitimate, there are other aspects of the story that exist as presented by the BBC and AJE. Nonetheless, CNNI have chosen to emphasise and advocate a terrorism discourse as an American broadcaster, which is an indication that CNNI's reporting is based on a topic which is high on the agenda in American politics. This is an interesting example since it showcases how global news networks tackle their political coverage through nationalistic interests and framing.

Interestingly, the first report analysed for that day by AJE was not that different from the BBC coverage, in the sense that it also stressed Johnston's personal experience. There was also more focus on Hamas's involvement and success in securing Johnston's release. The AJE report includes the quotes from Johnston on his experience:

Like being buried alive really, removed from the world and occasionally terrifying. I dreamt many times, literally dreamt of being free again and always woke up in that room. It's almost hard to believe that I'm not going to wake up in minutes in that room again, I don't think so the way things are going.

The choice to include this quote by Johnston, which includes the rhetorical tropes of 'buried alive' and 'removed from the world', reflects Johnston's negative experience during his captivity. However the remainder of the quote focuses on the optimistic moments of his kidnapping through the lexical items of 'dreamt of being free again' and 'I'm not going to wake up in minutes in that room again' This description highlights the feeling of relief that Johnston has, which is a positive one. So here the AJE report is generally descriptive, and although it includes a terrifying account of his kidnapping, it does include a quote which states that the terror is over. Although the personal account of Johnston's kidnapping is included in the reporting by the BBC and CNNI, AJE offers a similar account but clearly not as accentuated as the threat of being killed as in the CNNI report.

The AJE report subsequently shows Johnston hand in hand with the Palestinian leadership, before sitting down for a press conference. Here AJE highlights the role Hamas has played and proposes that, according to many, Hamas's involvement in Johnston's release was a bid to improve its image and gain international recognition. AJE states this explicitly: 'Many believe Alan's release is a sign that Hamas wants to start positively building its reputation internationally'. This proposition starts with the lexical items 'many believe' as an attribution of what is to follow. This reflects possibly either a general opinion or a constructed message where many journalists often attribute what they want to suggest to anonymous sources such as, in this instance, 'many believe'. Therefore, this presupposition, that assumes that Hamas wants to improve its international image, reads more as a suggestive claim than an attributed fact. Here AJE is suggesting that Johnston's release is proof that Hamas wants to improve its reputation globally. This is significant coming from AJE, as an Arab broadcaster, which would want to convey this message. Here, AJE does not use strong words such as the BBC World's lexical choice of 'shunned', rather it takes a more positive framing of how Johnston's kidnapping is an opportunity to build 'its reputation internationally'.

The AJE report also tells of the ordeal behind Johnston's kidnapping. It uses the lexical choice of 'masked gunmen' to describe the kidnappers. It is interesting to note that AJE here describes the kidnappers as simply 'masked' and 'gunmen' thus not over-textualizing the description of the kidnappers. Hence, this is quite a politically neutral description that does not carry much political connotation. The report then says that Johnston's kidnapping took him 'from the world he was passionate about' in reference to the Gaza Strip, thus emphasising that Johnston was emotionally interested in the Palestinian cause. At the same time the report is quick to point out that 'he knows all too well the risks on the streets of Gaza', meaning that what happened was not an unexpected scenario. One significant point that was not present in this report is that it does not include any footage of Johnston's tapes in the explosive jacket, nor any other online statements that were made by the kidnappers themselves, it merely describes them as 'masked gunmen' thus avoiding the terrorism discourse.

In this report it seems that AJE is careful to minimise the impact and description of the kidnappers, where it is attempting to avoid the terrorism discourse. It is interesting to note that through the same news story, one channel such as CNNI can accentuate a particular discourse such as terrorism, and another channel such as AJE attempts to minimise it. This provides an example of how media power in the construction of news discourses operates within a global arena, where discourses are a reflection of the national affiliation of the global news network.

A second report by AJE has been incorporated as part of this comparative analysis because it offers a slightly distinct discourse to the AJE report discussed above. This second report provides more background on Palestinian politics as a backdrop to the kidnapping, which would be more of interest to an Arab audience and hence would be the kind of reporting usually found in Arabic media. One possible explanation for this difference in reporting within AJE itself is that while the first AJE report is compiled by a Western reporter, the following report is presented by an Arab reporter. It may be the case that the Arab reporter's cultural identity and acculturation within an Arab news production environment contributes to this difference in emphasis. Examples of the difference between the AJE Western reporter and the AJE Arab reporter include the use of the Arabic name of the kidnapping group, Jeish Al Islam, by the Arab reporter meaning The Army of Islam. The reporter describes them as a 'fringe Palestinian group', which is quite a different description from that of the CNNI's overtextualisation of 'shadowy'. The word 'fringe' positions the group as operating on the edges, as opposed to 'shadowy' which has a more negative undertone. Moreover, although this description of 'fringe' is more politically neutral in description in comparison to the CNNI report, it does hold more of a political positioning by situating the group as operating on the edges than the AJE Western reporter who merely describes the kidnapping group as 'masked gunmen' as a description of their image. Such examples illustrate how discursive media power is constructed.

Another interesting example of how the two AJE reports differ is that the Arab reporter explains that the kidnappers are one of three groups that had claimed responsibility for the kidnapping of an Israeli corporal, Gilad Shalit. This explanation provides further insight as to what this group has been involved in, and offers more background about the kidnappers which is not provided by the other studied channels. In this instance, AJE is providing further background on the kidnappers. The reporter mentions the intervention of the 'Public Resistance Committees' which is the third group that was involved with Hamas in the kidnapping of the Israeli Corporal. The reporter then goes to talk further about a 'trilateral council of religious clerics' who met with the kidnapping group to discuss a potential release. This is a comprehensive narrative that serves as a background to the kidnapping which is not explained in the other studied reports. The report later provides an account of the Palestinian political environment in Gaza and states that the timing of the release coincides with the emergency government paying public sector employees their full wages after not being paid for seventeen months. It introduces this by saying: 'The timing of the release is also significant'. This proposition offers the viewer further insight into the political situation within Gaza, and thus connects Johnston's release to Palestinian discourses. The Arab AJE reporter has thus provided more detailed reporting on the story by linking the group to other ongoing Palestinian political stories, thus positioning the Johnston's kidnapping within an Arab discursive context. Such an example reflects that AJE's reporting does not necessarily constitute a 'challenge', rather AJE is offering insight to the kidnapping that would also be of interest to an Arab audience. In this sense, AJE can be regarded as 'informing' the audience about the contextual links that the Johnston kidnapping has within Palestinian politics, possibly for its own purposes or though its journalistic culture of reporting on the region. Therefore, we see in this report a deeper Arab focus on the coverage of the story which was not present in BBC and CNNI's reporting and, as this analysis exhibited, was also not found by the Western reporter in the first report within AJE itself. Although this report does not offer a 'challenge', it does reflect a form of 'informative' reporting into the kidnapping through an Arab reporting perspective and interest, and could also arguably be seen at times as adopting a public diplomacy role by emphasising Hamas's successful intervention in the case. Again although this discourse is not dictated by Hamas itself, AJE as an Arab broadcaster seems inclined to cover Hamas's role in the Johnston case.

In comparing all reports analysed on the Johnston case, it can be noted that the basic elements making up the story were included by all three channels. However what differs in these reports is the selection of what to include in the news reports; that is, the discourse that each report presents by focusing on aspects of the story that are considered important to each channel. Therefore, the three stages of the thematic structure have correspondingly revealed the following three points.

Firstly, there seems to be similar thematic items across the three channels studied. In the first step of the thematic structure, the deletion process involved the reduction of the news text by eliminating the thematic items of the news story that were parallel across all three channels. These items included in the reports provided, more or less, the same basic news items across the three channels. For example, how Johnston was kidnapped, who kidnapped him, threats made, pleas made, vigils organised, images of his release, quotes from Johnston, quotes from Hamas leadership, breakfast with Hamas, meeting the British consulate, and so on. Although it is not within the scope of this study to explore the sourcing of news, this analysis has revealed that global news networks can provide similar 'facts', however the discursive contexts vary.

Accordingly, this leads to the second point where although the 'factual' information is similar, the construction of the story varies from one channel to another. Thus the selection of which secondary information is to be included and omitted reveals the editorial decisions made during the making of the news reports. This editorial decision on what angle to pursue in the structure of the news reports reflects each channel's political economy be it a British, American or Arab representation. The BBC, as a British broadcaster and also an actor in the story, focused on the diplomatic relations between Britain and the Palestinian leadership, indicating that the kidnapping was not an official Palestinian position. This was evident in the use of government sources, the indication of the commonality of an editorial stance with the UK government, and the reciprocity of perspectives presented in report and quote. On the other hand, CNNI strongly focused on highlighting the terrorism discourse by making connections to the kidnappers and Al Qaeda, and which may also be 'indexed' to government agendas or appealing to popular concerns regarding terrorism. As for AJE, it was keen to emphasise that the diplomatic relations between Britain and the Palestinians were not tarnished, in addition to showing the good will of the Palestinian leadership and the efforts it went through to secure Johnston's release. So although there is an area where all three channels overlap in their reporting, there is a distinctive version of the story that each channel brings to the global audience and which was revealed through the thematic structure process.

With these similarities and differences established through the thematic structure, the main question in relation to this study remains: to what extent can AJE be seen as challenging the content of reports produced by BBC and CNNI? This question leads to the third and final point: AJE's reporting did not offer a clear 'challenge' in this instance, rather AJE's performance can be better described as providing a more in depth 'informative' discourse by including more background information and contextual links to the story. Yet there were times when AJE framed elements of the report differently, for example AJE described the kidnappers as 'masked gunmen' and 'fringe', whereas CNNI labelled them as 'shadowy' which holds a more negative connotation. These lexical items indicate the differences in how the kidnappers are perceived by each channel, and in turn how they are potentially perceived by their audiences. Such examples in the choice of words depict AJE as attempting to distinguish itself from Western global broadcasters. Such careful use of wording reflects AJE's concern to avoid dominant global representations, thus reporting on both the kidnappers and Hamas slightly more positively. Although such instances in this news story were small, they do reveal AJE's concern to offer a varying perspective. This variation from BBC World and CNNI, exemplifies how AJE is practicing its own nuance of media power where although its discourses are not that sharply contrasted to those of BBC and CNNI, there are elements of AJE's discourse (such the description of the kidnappers and of Hamas, the lexical choices, and contextualisations and links made) that reveal that AJE is covering the story through its Arab position.

Moreover, within AJE's reporting itself there were further discrepancies. The second AJE report analysed on the day of Johnston's release is an example where AJE has clearly adopted the Arab story in relation to Johnston's kidnapping. This report introduced additional information from an Arab news angle, such as the economic situation in Gaza and the role that Hamas as a new leadership is playing in this. This is not included in the BBC World or CNNI reports since these details might not fit with the angle these broadcasters report from, or might not even be of interest to their audiences. Yet, this representation by AJE cannot be regarded as a wholesale challenge to content found in BBC World and CNNI reports. AJE can be considered as offering more of what Arabic-speaking broadcasters would include in their reports, which is the focus on Hamas's role in Palestinian politics and the various factions involved behind the release of Johnston. And although no consistent contestation was found in AJE's news reports on the Johnston case, there were examples where its reports did offer moments of challenge, as illustrated in the labelling that AJE adopted when describing the kidnappers or Hamas. The next news story in this chapter will shed further light on these points.

In light of this analysis, it may be argued that AJE's news reporting on the Johnston case was not as radical as might have been expected, given both the way AJE has portrayed itself, and in light of claims made about it in academic debates. Claims made by AJE and others, that present it as a radical 'counter-balance' and 'contra-flow' to dominant international news media, were not matched in the findings of the analytic study conducted on the Johnston case. The reports by AJE that did introduce a different perspective to that of the other two Western broadcasters did not actually contest any information found in reports by BBC World and CNNI, rather, at most, they offered a different interpretive perspective into the same events. It may of course be argued that the provision of different perspectives is significant, not least where it is linked to differences in political perspective, and greater attention being paid to the politics of the Arab world itself. However, it is hard to argue that AJE's coverage provides a radical counter-narrative to the BBC World and CNNI reports.

From this first analysis it can be argued that there were a variety of journalistic influences that shape the discursive output of the studied channels, as discussed in Chapter 1. This relates to Cottle's (2003) argument, in Chapter 3, for the need to contextualise discourse analysis in relation to various influences beyond the text. The implications of this is the emergence of discursive outputs which in turn reflects the various journalistic influences which shape mediated discourse, such as the political economy of the news organisation, and the culture of the journalist, and the sourcing of news. These influences will be discussed in detail in the final chapter. First, however, since the analysis of one story provides a limited basis of evidence, the following story on the six medics who were sentenced to death in Libya sheds further light on this discussion in understanding AJE's news performance in its first year of operation.

The Medics and HIV

The previous analysis provided a first glimpse into the nature of AJE's reporting in comparison to BBC World and CNNI. The medics' case analysed here focuses on a political dispute which, in turn, reveals clearer instances of 'contestation' by AJE than those detected in the previous analysis. The medics' case includes parallel aspects to that of Johnston's story, where for example, it also contains a scenario where Westerners, working in the Arab world, are the centre of a political conflict and their lives are at risk; there is also a period of uncertainty where demands are being made; then there is an element of the story where diplomatic talks are taking place in the background, which eventually result in the release of the medics. However it is worth noting here, that the medics' case illustrates differences in reporting between AJE, BBC World and CNNI that are more accentuated than the Johnston case as there is a conspiracy theory involved that AJE brings out in its reporting, where it accuses the West of using the Arab world as a test Lab for the HIV/AIDS virus.

The setting to this story is that a team of six medical workers, comprised of five Bulgarian nurses and one Palestinian doctor, had been sentenced to death on the charge of infecting over 400 children with HIV/AIDS while working in a children's hospital in the city of Benghazi in Libya. The medics had spent eight and a half years in a Libyan jail before their eventual release. This news story also has various time lines which include the outbreak of the story, the death sentence, the appeal to the death sentence and then their release. The two milestones in the story that were included in the period of recording were the hearing of appeal to the death sentence which

Table 4.2 Common thematic items of the medics' story used by all three channels

Common Thematic Discourse Items
Medics accused of being guilty
Medics sentenced to death
Medics' appeal
Medics insist on their innocence
Images of courtroom
Over 400 children infected with HIV
Images of angry Libyan protestors
Images of Libyans holding pictures of infected children
Poor hygiene and sharing of needles behind spread of virus
Bulgaria says acceptance of compensation is admission of guilt
Death sentence upheld
Verdict might not be final word
A deal to be reached with families of infected children

began on 20 June 2007, and the upholding of the death sentence by the Libyan Court on 11 July 2007.

In conducting a comparative CDA on the reports surrounding the medics' case, similar observations to the Johnston case have been made. The thematic structure conducted revealed that all three studied news channels adopted a certain ideological stance on the story. These distinctions may, again, be seen to be related to the different cultural and ideological differences stemming from the social, cultural and political environments in which each news organisation is based. However, it is worth noting that, like the Johnston case, the main thematic items of the journalistic narrative were very similar in all reports, as it did not appear that one channel had more access to information than the other, as the above Table 4.2 outlines.

The discourses that emerge from the thematic structure reveal that all three channels are quite distinct in the discursive reporting of this story. BBC World is highly focused on the European position in relation to the case, and takes a one-sided stance with little coverage on the Libyan children and their families. CNNI, on the other hand, places emphasis on the Libyan children since its discourse is that of the AIDS virus, which is high on the agenda within American media. As for AJE, it stresses the suffering of the Libyan children; however it is keen to portray Libya as part of a conspiracy from the West on the Arab world.

On 20 June 2007, Libya's Supreme Court began hearing an appeal made by the medics. This follows a visit made by the European Union (EU) delegation to the nurses nine days earlier. Since the appeal hearing had started that day there was therefore little to report on, and thus the visit of the EU delegation was featured prominently that day. The BBC World report broadcast on this occasion introduces a background to the events thus far and points out that the medics insist they are innocent and that they confessed as a result of torture. It says: 'The five Bulgarian nurses and a Palestinian doctor – who always insisted that they're innocent – they say they've been tortured into making confessions'. This proposition is a form of modality which presents the nurses' innocence as a given truth. The clause, 'who always insisted that they're innocent', interrupts the proposition as background information here. It includes the lexical choices 'always' which indicates consistency and 'insisted' which reflects assertiveness, which are both strong words that serve to emphasise the nurses and doctor's innocence. The use of the lexical item 'innocent', which here is a presupposition, in addition to the inclusion of the word 'tortured', are lexical items that suggest that the nurses are not behind the infection of the children, and are seen as innocent. It also introduces them as being 'sentenced to death for allegedly infecting', where the presupposition 'allegedly' here suggests that, despite its guilty verdict, the ruling of the court has not satisfactorily resolved the allegations, suggesting to the viewer that a miscarriage of justice may be occurring. This introduction to the report is attempting to establish the innocence of the nurses, which is a clear indication of the position the BBC is adopting and advocating in relation to this story. These choices of words set the discursive direction for the report, which in turn provides an example of how lexical items can be regarded as constructs of discursive media power, where they frame the discourse towards a particular direction.

Figure 4.4   Image from the BBC of the European Union delegate with the nurses (BBC World, 20 June 2007)

The BBC report goes on to show the visit of the EU delegates to the nurses, with a focus on one female delegate who is wearing white and embracing the two nurses in assurance (see Image 4.4 above). The visual connotative effect of this almost gives the female delegate an angelic role. Thus, rather than criminalise the nurses the report implicitly proposes that the nurses are stuck in a situation where they cannot get out of and they need saving, as indicated by the proposition that the visit 'raised hopes' for the medics. This report provides many examples of how the verbal-visual correspondence plays a role in constructing a particular discourse. Again, it is worth noting that although the footage shown might indicate one aspect of the news story, the accompanying commentary frames the images in accordance with the news network's presented discourse. The use of images with the commentary is a significant method in the formation of discursive media power that is used in television news reporting, and in this case the BBC is doing this to build support in favour of the nurses.

The narrative of the above analysed BBC World report does not pay much attention to the infected children. Apart from the background images of the children which were used as a backdrop for the story's factual information, there is no mention of the infected Libyan children. Instead the report focuses on the medics and how they have been caught up in this problem. This report is surprisingly one-sided and this could be due to the fact that, as a member of the EU, the BBC is more interested to report on the story from a European angle. For example, the reporter states that 'For Libyan leader, Muammar Ghaddafi, the plight of the medics has become a real obstacle in his efforts to improve relations with Europe' which here is more of a presupposition that the medics' case is obstructing the Libyan leader's attempts to amend ties with Europe, since there are no specific facts to support this statement. Here the BBC, as a British broadcaster, is adopting a European position where its coverage is focusing on the Bulgarian medics as victims without much regards to the Libyan children as victims.

The AJE report broadcast that day is introduced against a backdrop of images from court. It starts with explaining the story thus far by saying that '[t]he five Bulgarian nurses and one Palestinian doctor were found guilty in 2004 of infecting children with HIV, a charge they deny'. The modality presented here through the lexical choices 'found guilty' and 'a charge they deny' seem to establish the guilt of the nurses, suggesting that the charge is effective however the nurses 'deny' the charge. These lexical choices are different to those found in the introduction of the BBC report which uses the word 'allegedly' in the proposition 'sentenced to death for allegedly infecting' which, through implication, establishes that the medics have been accused – another example of modality. The AJE report then starts by outlining the events of the case and points out that: 'The medics claim they're innocent, demanded that they were tortured to confessing.' It is notable here that this report uses, as lexical choices, the words 'innocent' and 'tortured' which are similar to the BBC report. Nonetheless, in the statement found in the BBC report which says: 'always insisted that they're innocent', the words 'always and 'insisted' as an example of modality are more assertive than the AJE report which uses the words 'claim'.

The report also includes a quote by US president (at the time) George Bush's speech where he demands the release of the nurses. We may note that AJE is the only news network to include the American President's quote, as it is common in Arab reporting to continuously include the American reaction to a story as a result of the US's close involvement in Middle Eastern politics. This is stated in the report as follows: 'And even President Bush said the release of the nurses was a high priority'. It is notable that the AJE report then shows the images of the infected children held by Libyan protestors and says that out of the 420 children, 52 have died. These figures reveal the severity of the situation. This is further emphasised through images of protestors holding posters with the following written on them: 'Libyan Kid's Cry for Children Human Wrights!' [original dictation]. This poster shows how AJE focuses on the children as being targeted. In this part of the report, AJE includes the number of children who have died, and focuses on a poster requesting the need for human rights considerations towards the children. Again, this is a slightly different coverage to the BBC which avoids talking about the children as victims. Accordingly, this AJE report does not present distinct discourses, rather it offers a narrative about the Bulgarian nurses and also focuses on the Libyan children. Although this report selects some lexical items that in someway lean towards establishing the guilt of the nurses, it is a mainly neutral report in comparison to other AJE reports that were broadcast on the day the court upheld the death sentence.

On 11 July 2007 the Libyan Supreme Court announced that the death sentence had been upheld following the appeal. The media, however, focus on talks of a deal that is underway, which could result in the release of the medics. The court decision is an awaited result, representing a milestone in the case, and therefore receives a reasonable amount of coverage across all three channels. On this day the overall coverage on CNNI includes the mapping of events related to this story and the interviewing of various experts on Libyan politics. The report begins with the news anchor stating that the medics went to Libya to save the lives of children but instead found themselves accused of infecting the Libyan children with AIDS:

'They went to Libya to save lives but three years later they were sentenced to death convicted of purposely infecting hundreds of Libyan Children with the AIDS virus.' Here the news anchor establishes from the start of the story, as a form of modality, the innocence of the nurses as a given truth by providing a judgement on their intentions when saying they went to Libya wanting to save lives, which is different from the AJE report that day which states that the Libyan children were targeted, as I shall go on to discuss. This first clause of the proposition, in the CNNI report, is narrated as a factual background to the medics' story. This statement of good intentions is a presupposition which serves to frame the story from the outset, which might be deemed problematic since their intent is an issue that is still under trial and it is therefore questionable how the news anchor could establish this. An implication is also deployed where this statement proposes that the nurses found themselves 'convicted'. The introduction by the news anchor concludes by explaining the procedure of the case and how there may still be hope for the nurses' release. The news anchor's use of language offers a framing to the story which presumes the nurses' innocence. This introduction to the story establishes the nurses as victims from the start of the report, instead of just telling the story as it is. The way in which the report is introduced reflects how CNNI is exercising discursive media power by framing the nurses as innocent, where CNNI is taking a 'position' on the story.

Interestingly, however, the report itself does not follow the anchor's introduction that emphasises the nurses and their plea of innocence. Rather, it provides many images of the deceased and infected children. It is worth pointing out that CNNI is the only channel that shows images of deceased children out of all the reports studied, as the BBC only includes images of the infected children on a background poster and AJE only includes images of infected children in the hospital. This CNNI report shows various images of the infected children in hospital wards. The CNNI reporter starts by saying that 50 children have thus far died out of 400 Libyan children: 'Five Bulgarian nurses and a Palestinian doctor found guilty of deliberately infecting more than 400 Libyan children under their care with the virus that causes AIDS, 50 children have died'. The words 'under their care' imply that some responsibility is placed on the nurses. On the other hand, this report also offers some insight into the nurses' situation by also being the only report out of the studied data to include a quote from one of the nurses saying she is innocent: 'No matter what people say, I'm simply innocent, that's all.' Surprisingly, this attention to the children at the start of the report is very different to the construction of the BBC report or even the AJE report in the first event. It is noticeably sympathetic towards the children, whilst also presenting the nurses' side of the story. This continues throughout the report.

Contrary to the CNNI report, the BBC report broadcast that day again did not include the children as part of the story, but mainly focused on the political talks surrounding the case. This report starts with images of the judges calling out the names of the nurses, and talks about how despite the death sentence being upheld a deal might be underway. The report then mentions the relatives of the nurses and includes images of them holding vigils at the Libyan Embassy in Sofia. This is the only instance in the reports studied where there is mention of the nurses' families: 'Keeping vigils outside the Libyan Embassy in the Bulgarian capital, relatives of the nurses are angry and distressed'. This proposition includes the lexical items 'angry and distressed' to describe the feelings of the nurses' relatives, thus showcasing the frustration of the nurses' families. This is accompanied by images of the nurses on the phone looking worried. This report here identifies the nurses as the victims. This section of the report is the only example from the data collected that focuses on the families of the nurses, by using emotional images and language to portray their agony and helplessness. Yet throughout this report there is no mention of the children, which is an indication that this report is advocating the plight of the medics. This is further emphasised through the reporter's choice of lexical items to establish the nurses' innocence.

The BBC report shows the Libyan leader with European officials, and says: 'Since abandoning his pursuit of weapons of mass destruction, the Libyan leader, Colonel Ghaddafi, has brought his country back in from the cold'. The first clause frames the Libyan leader negatively by showcasing Libya here as a state that has been anti-Western by wanting to develop 'weapons of mass destruction'. The report then ends with saying that solving this case could be 'the last obstacle to Libya's full rehabilitation'. This presupposition, that the medics' case was the 'last' obstacle in amending its ties with the Western world, adopts the lexical choice of 'rehabilitation' here to depict Libya – at the time of the broadcast of this report – as a country that is not yet approved of by the international community after just having come out of lengthy sanctions as a result of the Lockerbie bombings. Here the report is including the Libyan leadership's history which reminds the viewer of the negative ties the Libyan leadership has historically had with the international community and that this case is an 'obstacle' to Libya's 'rehabilitation'. The use of these words indicates a discourse that is not favouring Libya, and which in turn enforce the BBC's discourse of supporting the Bulgarian nurses and adopting a European/Western position. Here the choice of words again illustrate how lexical items can be effective constructs of discursive media power, where such lexical choices help frame the story towards a particular discourse.

It is interesting to point out at this stage that, overall, the reports studied in the medics' case do not include any details of the evidence used in court that has led to the outcome of the death sentence. Hence, speculations continue in the media reports without any facts to assist the viewer in forming an independent opinion or taking a position. This is further exemplified in the following AJE report which also offers speculations to assist its argument.

Contrary to the BBC's coverage, AJE reported on this story with great attention to the children, similar to CNNI. The following report by AJE is included here since it focuses on this point and in turn provides an example of AJE to some degree includes contesting views to the above analysed reports by the BBC and CNNI. What is significant is that this report focuses on the children as victims – not on the medics – placing the emphasis on the children, and thus weighing the issue rather differently.

Table 4.3 AJE, 11 July 2007: 'Libya AIDS trial: Verdict due Wednesday for foreign medics' (04.10 length)

Text	Image
[Intro] Now a Libyan court is expected to rule on an appeal by six foreign medics who are on death row after being convicted of infecting children with HIV. In a development ahead of that ruling though, the Ghaddafi Foundation has said it's reached a compensation deal with the families of the children which could pave the way for the release of the five Bulgarians and one Palestinian.	Background behind the anchor is a yellowish photo of children in the hospital blended in with photos of the nurses in the courtroom behind bars. A veiled Libyan mother caressing her daughter on a hospital bed, followed by the camera focusing on the mother's hand which is decorated with Henna stroking her daughter's hand, which is also decorated with Henna, the mother's decorated hand goes on to stroke her daughter's face.
Nora Taha has lived all her ten years with HIV. She's one of more than 400 children. It's alleged to have been deliberately infected with the virus by the Palestinian doctor and his team of Bulgarian nurses.	The mother speaking, with the captions: 'Salma Taha, Nora's Mother'.
'When we're invited to weddings or dinners, people point at us. "That's the HIV positive girl and that's her mother", they say. We're told that she can't eat from this plate or drink from that cup. We're isolated and devastated'.	Nora's mother's head leaning on Nora's head, then the mother lifts her head and tears are running down her face.
Nora's life now depends on taking a toxic cocktail of lethal viral drugs.	A shot taken from the end of the bed of Nora's mother caressing Nora in bed while stroking her daughter's hair and holding her chin. Then two boys in a hospital ward on their beds, then the camera moves to another

All the children here and their parents believe they were victims of an experiment to find a cure for HIV. It seems an incredible claim and one that goes against international scientific opinion on the case. But here they insist that each victim was injected with the same strain of the virus. If it had been accidental, the virus would have infected from person to person.

'The motive in this case, maybe they were misguided; those people were misguided by a third party that convinced them that they're part of a great scientific experiment to finding a vaccine or something, that horribly went wrong'.

No one claims to know what organisation is behind all of this but suggestions are firms desperate to find a cure for HIV.

So why choose this Mediterranean city for the so called medical experiment? Well, with little travel in and out of the country patients were an untainted sample and if things go wrong, the eccentric leadership could be blamed. That any of these victims is willing to speak about what's happened to them, is extremely rare in the Arab world.

bed in the room where a boy is lying down with a sad face and looking into the camera from far away. The camera then takes a very close shot of the boy's face while he is looking away. This is followed by the image of a doctor wearing white plastic gloves and feeling the boy's stomach with the white gloves.

A man wearing a shirt and glasses sitting in the ward making the comment with caption: 'Dr Imran Al-Factory, Director of Contagious Diseases Centre.'

Men and women walking out of doors and into corridors of a building, then an image of the building from the outside with people walking outside it and zooming out then focusing on the name of the building in Arabic which is the Contagious Diseases Centre.

A skyline of the city from a highway with cars passing by, this is followed with a shot of the city with water, palm trees and buildings that zooms into more buildings in the background. The reporter, speaking to the camera making a point by using hand gestures; he is on the beach with water and sand in the background against an oriental looking building.

But medical procedures at the hospital where the children were infected have been condemned by the French doctor who discovered HIV. In a report commissioned by the Libyans, Dr Luc Montagnier blames dirty needles for spreading the virus. He says the outbreak almost certainly began with a single child infected by his mother and then spread, but it wasn't intentional.

That latest report may be presented to the court for Wednesday's hearing, but if the doctor and the nurses are eventually freed it will be down more to international politics and compensation deals rather than whether they are actually guilty or not.

A shot of both the same boy and girl included in the previous report, but this time focusing first with a close-up on the boy's face and then on the girl's face. This is followed by the image of the boy and girl each in their hospital beds, and the back of a woman looking outside the window and the girl looking at her. At that same shot the camera then moves to the end of the two beds where two women are sitting, one worried and the other crying. The camera then focuses on the face of the woman crying who has covered her eyes with her head scarf while wiping her tear. The next shot shows the same woman's face with red eyes crying and talking however the viewer cannot hear her voice as there are only images that are accompanying the reporter's comments. This shot ends with the camera zooming out of two animation characters painted on the wall next the two beds with the boy and girl in them. Back to the panel of judges in the courtroom, with various shots from previous footage of the medics in the cage in the courtroom, and then one of the nurses being guided to the cage by another nurse, and ending with the shot of the Palestinian doctor sitting in the cage smiling and talking.

In the AJE report included here, the news anchor introduces the story by describing 'six foreign medics who are on death row after being convicted of infecting children with HIV'. Note that the last clause of this proposition uses the lexical items 'convicted of infecting' which, although is the outcome of the court rulings, is different from the introduction of the BBC World report which uses the words 'allegedly infecting', or CNNI's implication of innocence in the statement 'They went to Libya to save lives'. The modality used by both the BBC World and CNNI, thus, introduce their reports with a suggestion that the medics are innocent. While, in each case, the nature of the story is given definition through particular forms of presupposition, it is particularly notable that only AJE appears to presuppose the guilt of the medics by presenting the conviction as legitimate. This reveals how the choice of lexical items again reflects nuances of discursive media power, thus exposing the discursive choice of news networks.

The above AJE report starts with a specific narration of one of the Libyan children's case who is infected in hospital with HIV, a journalistic convention that is used to personify the issue and bring it closer to the audience by humanising the problem. The reporter introduces the story as follows: 'Nora Taha has lived all her ten years with HIV. She's one of more than 400 children.' The inclusion of the second clause in this proposition serves to highlight the significance of Nora's story since she is not the only case but rather one out of many. He goes on to say that the case of Nora is 'alleged to have been deliberately infected'. Although the reporter uses the lexical item 'alleged', which does not hold any confirmation to the infection, it also uses the word 'deliberately', which again enforces the notion of it being an intended operation. The report tells her story while showing detailed emotional images with her mother in the hospital (see Image 4.5). It also includes a quote from Nora's mother who talks about how HIV has isolated them from society and how they are treated differently: 'When we're invited to weddings or dinners, people point at us. "That's the HIV positive girl and that's her mother", they say. We're told that she can't eat from this plate or drink from that cup. We're isolated and devastated'. The inclusion of

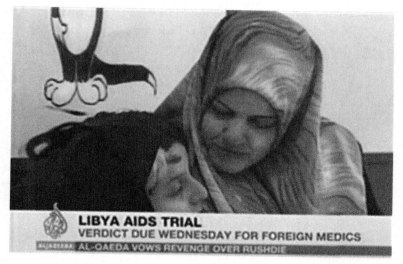

Figure 4.5   Image from AJE of Nora Taha and her mother in hospital (AJE, 11 July 2007)

the lexical items 'isolated and devastated' from this quote, serve to emphasise the plight of the children and their families. These propositions provide an emotional angle to the suffering the children and their parents are enduring, and details the agony they are going through as a result of the social stigma associated with the disease. The reporter then talks about the medications Nora is taking while images are being played of the mother crying with her head leaning against her daughter's head: 'Nora's life now depends on taking a toxic cocktail of lethal viral drugs'. The lexical choices of 'toxic cocktail' and 'lethal' here by the reporter, provoke pity and anger for what the patient has to go through as a result of the case.

Nora's story here is used dramatically by the reporter to emphasise the children's suffering, and takes up a substantial part of the report. The use of emotional appeals, as used in the BBC's report when showing the families of the nurses outside the Libyan Embassy in Sofia, is also used in this AJE report by bringing the suffering of Nora and her mother to the international audience. In both these reports, the use of emotional appeals is used differently, thus each constructing a distinct application of discursive media power: The BBC in support of the nurses and AJE in support of the children.

This tone and style of coverage in AJE's report continues. The AJE report moves on to show lengthy repetitive and detailed images of other infected children in hospital. The reporter states: 'All the children here and their parents believe they were victims of an experiment to find a cure for HIV'. The transitivity participants here are articulated through the lexical choice 'all' which places emphasis that this belief is unanimous, and thus cannot be doubted. The reporter go on to say that '[i]t seems an incredible claim and one that goes against international scientific opinion on the case. But here they insist that each victim was injected with the same strain of the virus'. In this proposition, the reporter uses the lexical choices 'it seems an incredible claim' in the first clause suggesting that it might not be realistic to assume that the children were victims of an experiment, however the reporter then goes on in the second clause to support the 'claim' with an unattributed argument that '[i]f it had been accidental, the virus would have infected from person to person'. This example of modality, which presents the children's deliberate infection as a given truth, is the reporter's personal rhetoric which supports the infected children and their parents' position, thus revealing that the reporter himself is taking a strong position against the nurses.

The reporter takes this further by quoting Dr Emran al-Faitouri, who is Director of Contagious Diseases Centre in Libya, as saying that the infection was a conspiracy: 'The motive in this case, maybe they were misguided; those people were misguided by a third party that convinced them that they're part of a great scientific experiment to finding a vaccine or something, that horribly went wrong.' This is the first conspiracy theory in this report. However, the AJE reporter then goes on to say that: 'No one claims to know what organisation is behind all of this but suggestions are firms desperate to find a cure for HIV.' From the transitivity process the transitivity participants are not clearly identified in this case as can be seen through the words 'what organisation', but are presupposed to be 'firms desperate to find a cure for HIV'. Again the reporter reiterates claims that are not backed by any factual evidence. The only support he provides to this is that '[n]o one claims to know' and 'suggestions', which is presented as a rhetorical credence through this repetition. This information is presented without any evidence, and although a doctor holding the position of Director for Contagious Diseases makes this claim, it is not one that is based on any scientific findings or court evidence. It is worth pointing out here that an expert is called on to support the discourse presented by the reporter without providing another expert to showcase the other side. This is another example of how news networks include particular sources, without the inclusion of the other side of the story, to enforce their presented discourse.

The second conspiracy theory is then introduced claiming that this is a Western experiment targeting an Arab city: 'So why choose this Mediterranean city for the so called medical experiment?'. This proposition serves as a form of modality where the reporter is establishing as a given truth that Benghazi was in fact chosen for the HIV experiment. In the first clause, the reporter is referring to the city of Benghazi, taking the argument further to implicate that it was a 'targeted' city. It is interesting to note here that the reporter uses the lexical choices 'so called medical experiment' where the reporter here is careful not to establish that it is a 'medical experiment' by including the words 'so called' before it, yet by implication this proposition serves to support the claim that the infection of the children is an experiment. He then elaborates further on this by stating that whilst emphasising that if anything were to go wrong in this experiment all could be blamed on the Libyan leadership because the Libyan leadership is seen by the Western world as an 'eccentric leadership'. The reporter then explains the reason behind the victims not speaking out. He explains that it is due to Arab culture and tradition that the victims and their families not speak up: 'That any of these victims is willing to speak about what's happened to them, is extremely rare in the Arab world.' Here it seems that the reporter needs to defend any gap that could be identified in his argument. It is also important to point out that the delivery of the AJE reporter here is highly dramatic with very expressive facial expressions, hand gestures and vocal tones (Lorenzo-Dus, 2009).

Further images of the infected children with their families crying are included (see detailed description of images in the report included above). The AJE report then goes back to the courtroom and concludes with a statement saying that the freedom of the medics will be more a result of 'international politics and compensation deals rather than whether they are actually guilty or not'. This presupposition is an implication that the medics will get away with their crime due to international pressures. In other words the reporter sees that the medics as Westerners have the backing of their government which is exerting international diplomatic pressure on Libya for their release. The AJE reporter is arguing towards a conspiracy by the West against an Arab nation, by providing explanations and applying rhetoric. The dramatic method of delivery is also used to accentuate his argument.

Overall, this is a significant report that has been included here in detail, as it provides an example of AJE's reporting that offers instances of 'contestation' compared to the reporting presented by the BBC and CNNI. This discourse reflects the political view of the Arab street where people constantly feel that they are being targeted by the West, as explained in Chapter 2. This in turn is reflected in Arab reporting where the discourses behind various issues are constantly being put forward in such a manner. This report did not hold much evidence. All points made against the medics were not taken from the court hearings or any other official statements, rather they were assumptions and implications made by the reporter himself. The reporter takes on quite a passionate and partial tone, and portrays the Arab city of Benghazi as one that is targeted by the West. The report focuses on the children as victims and the politics surrounding the case.

Again, it may be notable here that the reporter, is an Arab reporter, particularly as he adopts a reporting style and delivery that is very similar to that found in Arab media reporting. By contrast to a measured, 'impartial' stance associated with the tradition of journalistic 'objectivity', his reporting style is extremely passionate and emotional. The differences in reporting between this report and those by BBC, CNNI and even the AJE report analysed which was compiled by a Western reporter, is significant. It is clearly distinct from discourses found in the other reports, especially the BBC reports which tended to avoid mentioning the Libyan children and rather aimed at focusing on the medics. This, again, provides a second case in which a stronger level of difference between AJE's report and those of its competitors coincides with its presentation by an Arab reporter, raising the question of whether the identity of journalists is a significant factor in shaping the form of journalism that is presented in AJE's output. Overall, this coverage of the medics' case has provided an example of how AJE attempts to somewhat 'contest' the discourses of CNNI and the BBC, and is doing so through the application of its selected discourse.

In conducting a comparative CDA on the reports surrounding the medics' case, similar observations to the Johnston case have been made in this chapter. The thematic structure conducted in this chapter revealed that all three studied news channels adopted a certain ideological stance on the story. These distinctions may, again, be seen to be related to the different cultural and ideological differences stemming from the social, cultural and political environments in which each news organisation is based. However, it is worth noting that, like the Johnston case, the main thematic items of the journalistic narrative were very similar in all reports, as it did not appear that one channel had more access to information than the other. The study of AJE reports in this chapter did not show that AJE sourced any independent news. For example, it would have been interesting, and even expected from AJE, to gain an understanding on some background information relating to the Palestinian doctor. AJE did not include any additional facts surrounding the Palestinian doctor, they did not interview his family, nor did they provide an account of his background. As an Arab broadcaster, AJE could have explained how a Palestinian doctor was involved in this claimed conspiracy from the West against an Arab country. They could have also provided some insight into any diplomatic attempts to secure his release within the Arab region. In other words, there was no sourcing and inclusion of information regarding the Palestinian doctor from AJE that was different from BBC World or CNNI. Although the

Palestinian doctor was, in fact, a Bulgarian citizen, the BBC reports focused on the Bulgarian nurses only.

In relation to the question on whether these results position AJE as a 'counter-balance' channel or not, it is worth pointing out that although the final report by AJE offered a counter-discourse to that of the BBC, this counter-position was not based on alternative evidence. The Arab reporter advocated the plight of the children to support the argument that Libya was a targeted country by the West without much factual proof. This report did not source any information as a background to the report and used one expert quote that was supportive of its line of reporting, and which did not hold any scientific evidence. Such form of reporting reflects a strong notion of challenge. In this sense, while it may be argued that this AJE report was not factually based, it does counter-balance the BBC report. The BBC report particularly focuses on the plight of the nurses, their families, their release and the political diplomatic attempts to secure their release. It frames Libya as a rogue state that is torturing the nurses as innocent victims. Contrary to this, AJE focuses on the plight of the Libyan children and suggests that Libya is a country targeted by Western companies that aim to carry out an experiment to find a cure for the AIDS virus, and has found in Libya a suitable target due to its 'eccentric leadership'. This discourse, therefore, is clearly a counter-argument when compared to the BBC report. However, it remains hard to argue, in its overall coverage of the medics' case, that AJE has 'contested media power' in a strong sense, rather AJE's news output can be better understood as a hybrid practice.

Through a CDA of AJE's news coverage in its first year of operation, this chapter uncovers and argues that a consideration of AJE as a hybrid broadcaster reflects various local and global forces that influence its performance. This sheds light on the complex nature of the field of satellite television news and highlights a need to move beyond binary frameworks of either homogenisation or hetrogenisation when considering developments in global communi-cation. AJE's performance at the start of its launch can be regarded as an example of Arjun Appadurai's description of global culture where 'sameness' and 'difference' blend to produce the 'triumphantly universal and the resiliently particular' (Appadurai, 1990: 307–308). The findings of this study have identified examples of news reporting that reflected 'sameness' and 'difference' simultaneously between television news networks. This study, therefore, adopts a 'domination and resistance' position on the question of a mediated 'global public sphere', by acknowledging the persistence of inequalities in flows and definitions of global satellite news whilst at the same time acknowledging the significance of emerging operations of broadcasters which represent different parts of the world. It suggests that a continued 'global dominance' and an expansion of a 'global public sphere' may be occurring simultaneously in the field of international news, and that AJE exemplifies and provides a vehicle for both processes. This point will be discussed in further detail in Chapter 6, however it is first necessary to re-visit AJE's textual output eight years later: post-Arab Spring.

CHAPTER 5

POST-ARAB SPRING

The reason behind identifying the Arab Spring protests as a landmark for re-visiting Al Jazeera English's (AJE) discursive practice in this book, is arguably due to the significance that these protests have had in Arab democratic processes, and in turn the plethora of Arab and global news coverage surrounding them. Indeed, the Arab protests marked the opening up of the Arab public sphere both on the streets and through the media, specifically social media. Within such an environment, the need to question the role of AJE throughout these protests becomes vital, given Al Jazeera has historically played a significant role in opening up the Arab communicative space as discussed in Chapter 2. The analysis conducted in Chapter 4 showed that AJE was somewhat hybrid in its journalistic practice within its first year of operation. The analysis indicated that the reporting of all three channels studied was generally dependent on the nature of the events that were reported on, and the national stance these broadcasters adopted in relation to these events. This chapter will look into whether these findings were replicated through events during the Arab Spring period, with a specific focus on the role of AJE. The events for analysis in this chapter are: the sentencing of 528 members of the Muslim Brotherhood on 24 March 2014 for participating in an attack on a police station; and the second event is the United Nations (UN) Security Council's meeting on 19 July 2012 which failed to adopt a UN resolution on Syria and which was vetoed by the Russian Federation and China. It is necessary first, however, to provide a background on the role Al Jazeera played within the context of the Arab Spring uprisings through existing literature and media reports.

The Arab Spring and Al Jazeera

The Arab Spring protests, as they came to be known, began in December 2010 and were instigated by a Tunisian street vendor by the name of Mohamed Bouazizi who eventually ended his life through self-immolation. On 17 December 2010, Bouazizi poured gasoline on himself and set himself on fire in front of the governor's office, an hour after his wares were confiscated by a female municipal inspector. The Tunisian revolution, or the 'Basta' revolution as it was called in the Arab world in reference to the fruit cart that Bouazizi used, marked the start of protests which spread through Tunisia and which lead to the ousting of the Tunisian President Zine El Abidine Ben Ali who fled to Saudi Arabia. The protests on the Tunisian streets poured over to other Arab countries and ultimately led to the cessation of other leaderships such as Egypt's Hosni Mubarak, Libya's Muammar Ghaddafi and Yemen's Ali Abdullah Saleh.

To understand the role of Al Jazeera during the Arab Spring protests and the discourses surrounding its involvement, it is necessary to gain an idea of Qatar's political involvement in the region during this time. During the events of the Arab Spring Qatar played an active political part reflecting 'a drastic shift away from its previous focus on diplomatic mediation in favour of actual intervention and picking sides in regional conflicts' (Ulrichsen, 2012: 12). Examples of Qatar's active role in the region include: joining NATO forces during the Libyan uprising 'to dilute concerns of another Western intervention in the region' (Ulrichsen, 2012: 13); championing the Muslim Brotherhood rule during the Egyptian elections; and supporting the Syrian opposition, calling for Syria's president, Hafez Al Assad, to step down during sessions at the Arab League and UN (Ulrichsen, 2012: 13–15). As a result of this involvement, Al Jazeera was banned in many Arab countries during the uprisings such as Libya where their office was closed immediately after the start of the revolution, and in Egypt where its satellite signal covering Egypt was dropped from NileSat (Khanfar, 2012). According to Figenschou, AJE's role in the Arab Spring protests was subject to 'structural limitations and contradictions' that have shaped the channel itself; she argues that 'the channel's resource-intensive production strategies make it highly dependent on its Qatari sponser' (Figenschou, 2013: 161). This explains the direction AJE adopted in its reporting on the protests, where the channel itself took a political part. This, in turn, enhanced its political power through its mediated discourses.

As a result of this, many prominent staff members left Al Jazeera following the Arab Spring protests. One correspondent said the reason behind his resignation was that 'Al-Jazeera takes a clear position in every country from which it reports – not based on journalistic priorities, but rather on the interests of the Foreign Ministry of Qatar' (Kühn et al., 2013). This supports the findings from the analysis conducted in Chapter 4 where each news network reported in accordance with its national political agenda. Whether this is positive or negative is not the point of discussion in this book, rather what these findings suggest in relation to the question this book poses, is that the reporting of international news networks within the global news space is dependent on various factors such as the actual event and the position the news channel takes in relation to this event.

Examples of Al Jazeera's political reporting during the Arab Spring include Al Jazeera's role in shaping discourses on Libya where at the start of the protests against Ghaddafi, the station began to use the rebel's tricolour flag, instead of Libya's official green flag (Baker, 2011). Qatar's direct involvement in the Arab Spring protests sought criticism especially when this political involvement was reflected in its reporting of Al Jazeera, as can be seen in the following example:

> Qatar's reputation also was dented by accusations that al-Jazeera English bowed to political pressure to not rebroadcast Shouting in the Dark, its award-winning documentary about

the Bahrain uprising. Shortly afterwards, in September 2011, al-Jazeera's longlasting Director-General, Wadah Khanfar, resigned suddenly, and was replaced by a ruling family member, Sheikh Ahmad bin Jassim Mohammed al-Thani' (Ulrichsen, 2012: 14).

The political role of Al Jazeera was also noted during the Syrian crisis, where Al Jazeera staff were reported as being very close to the rebels, which in turn was also reflected in Al Jazeera's reporting (Kühn et al., 2013). When the Arab League submitted a proposal to the Syrian government in November 2011 to address the crisis, the Syrian foreign minister replied 'that the deal should include a stipulation requiring "certain television channels" stop their "poisonous" reporting' (Souaiaia, 2011). Such examples note the power that news networks have in not only shaping public opinion, and in turn social reality, but also in becoming active political players themselves in the conflict. Ali Hashem, an Al Jazeera correspondent, spoke about his experience reporting on the Syrian revolution:

> I was one of those who experienced it when al-Jazeera, the channel I used to work for, refused to air footage of gunmen fighting the Syrian regime on the borders between Lebanon and Syria. I saw tens of gunmen crossing the borders in May last year – clear evidence that the Syrian revolution was becoming militarised. This didn't fit the required narrative of a clean and peaceful uprising, and so my seniors asked me to forget about gunmen (Hashem, 2012).

Al Jazeera's editorial policy was also evident in the case of Egypt. AJE's criticism of the protests against Mohammad Morsi and the Muslim Brotherhood reflected the editorial stance it took in regards to the politics of Egypt (Kühn et al., 2013). Ghaffar Hussain states in relation to this:

> Since the Muslim Brotherhood has come to power in Egypt, Al-Jazeera has done all in its power to portray the group in a

favourable light. Protests against the Brotherhood-dominated regime are presented as being led by violent thugs with no political grievances, while Morsi's poorly constructed and shallow speeches are given positive coverage. Similarly, the network has largely ignored the protests in neighbouring Bahrain while at the same time dedicating a great deal of attention to the protests in Syria (Hussain, 2013).

As a result of Al Jazeera's direct involvement in the political framing of the events post-Mubarak regime, several Al Jazeera reporters were imprisoned. These journalists included Australian correspondent Peter Greste, and producers Mohamed Fahmy and Baher Mohamed, who were arrested in Cairo on 29 December 2013, and who were accused of spreading news and belonging to the outlawed Muslim Brotherhood. Another AJE reporter, Abdullah Elshamy, had been detained without charge from 14 August 2013 till 18 June 2014. In an interview with Wadah Khanfar, who was director general of Al Jazeera network from 2003–2011, Hayley Sweetland Edwards said that '[w]hile Khanfar said Al Jazeera does not aspire to political activism ("We are journalists; we are not political activists!"), he acknowledged "participating in creating the environment for the Arab Spring"' (Edwards, 2011). What this means, is that Al Jazeera journalists were, to some degree, not just mediators but actors in the conflict, which emphasises the media power a news network can hold in shaping the political uprisings and the aftermath. A study conducted by the Regional Centre for Strategic Studies (RCSS), which is an independent think tank in Cairo, found that Al Jazeera Arabic (AJA) had lost its popularity with its audiences in the region as a result of Al Jazeera's biased coverage over Egypt's political happenings, whereby AJE would only invite the Muslim Brotherhood representatives to its programmes and disregard other political groups in Egypt (Youssef, 2013).

As noted in Chapter 2, prior to AJE's launch, discussions focused on whether AJA would be as alternative as its parent station AJA, or whether it would be a watered-down version of it. In the wake of the Arab Spring protests, discussions changed to whether

AJE followed the politically active role that AJA adopted in the region or whether AJE had its distinct line of reporting. From the coverage and discourses surrounding the role of AJE globally, there was a notion that AJE's popularity during the Arab Spring protests in fact rose due to it adopting a different reporting agenda to that of AJA. On this, Kristian Coates Ulrichsen argues that during the 'reporting of Egypt's 18-day "revolution" from Cairo's Tahrir Square [...] [i]t's live-streaming of the massive demonstrations against Hosni Mubarak resulted in a 2,500% rise in viewing figures and a growing clamour for the channel to be included on satellite television packages in the United States' (Ulrichsen, 2012: 11). However, when AJE began to broadcast to 2 million cable subscribers in New York on August 2011, Western audiences were expecting to get a 'direct or approximate translation of its parent network in Qatar' (Kessler, 2012). Yet as Oren Kessler notes: 'to appreciate what Al Jazeera English is, it is critical to remember just what it is not – even a remote likeness of its Arabic-speaking progenitor' (Kessler, 2012). Nevertheless, '[i]t seemed, al-Jazeera had won for itself international acceptance and credibility as its English-language channel succeeded in rebranding the network, even though its Arab-language channel remained significantly different in both content and tone towards the unfolding upheaval' (Ulrichsen, 2012: 11). This is an example of how AJE can be regarded as a 'subaltern counterpublic' (Fraser, 1993), seeking to gain access to a wider public.

It is interesting to note that when AJE was created, it was developed as an English version of AJA. Prior to AJE's launch, the editor-in-chief of AJA said that there is a need for an English channel for those who do not understand Arabic:

> I met so many people who watch Al Jazeera [...] People used to tell us, 'If only we could understand Arabic. We watch you for the pictures.' [...] I think people sometimes just see the pictures and they are waiting for Al Jazeera English to be launched' (Khamis, 2007: 46).

Yet, after the launch of AJE, journalists working at the channel emphasised the difference between the two channels. A former AJE correspondent in Cairo, Ayman Mohyeldin, commented on the difference between AJE and AJA saying: '[a]t the end of the day, we don't share the same editorial policies [. . .] What we do share is the editorial code of ethics and the same editorial vision as the network' (Kessler, 2012). Shawn Powers and Eytan Gilboa (2007) argue that Al Jazeera as an organisation has two discreet roles, one which is internal, and the other external. They describe these roles as follows:

> The internal is exemplified by the network's initiation of discussion on controversial and taboo topics in the Arab and Muslim public sphere, as well as by its continued scrutiny of Arab regimes. The external role is exemplified by Al Jazeera's claim to represent to the world Arab and Muslim perspectives on regional and international events (Powers and Gilboa, 2007: 53).

While it is not within the scope of this study to compare AJA's reporting to that of AJE, as other studies noted above have already done this, these debates emphasise the hybrid role within Al Jazeera through AJE and AJA, and which was even noted in Chapter 4 within AJE itself, where Arab reporters and Western reporters within AJE were found to adopt different discourses in line with their culture of journalism. This hybridity within AJE's discursive performance has also been noted in the two news stories analysed in this chapter: Egypt mass death sentences, and the veto of the UN resolution on Syria by Russia and China.

Egypt Mass Death Sentences

On 24 March 2014 Egypt's criminal court in Minya sentenced 528 members of the Muslim Brotherhood for participating in an attack on a police station in mid-August 2013, in which a member of the police force was killed. This happened in response to the security force's break-up of two demonstrations in support of ousted president, Mohammed Morsi in Cairo that resulted in 1,000 dead. What is particularly significant in this court ruling is that it was the

Table 5.1   Common thematic items of the Egypt death sentences story used by all three channels

Common Thematic Discourse Items

Largest number of people to be sentenced to death in single trial
Around 528 were sentenced to death
Around 398 were tried in absentia
Convicted of the murder of a police officer
Convicted of attempting to murder two police officers
Convicted of attacking a police station in Minya
Defence lawyers not given enough time to review documents
Defence lawyers were not allowed in the courtroom

largest mass ruling in a single trial in Egypt's modern history, where most of the defendants were also tried in absentia. The court did not allow the defence lawyers into the courtroom. Through a study of the news stories on the day of the court ruling, the common thematic themes by all three channels are outlined in Table 5.1.

Table 5.1 illustrates how the main components of the story were shared as the common narrative by all three channels. Again from this exercise, it can be noted that the table holds factual information that is necessary for conveying the story. While the above items represent the 'pillars' of the story that provide a complete picture of the event, each channel offered a different construction of the event which reflects the editorial decisions made. As discussed at the start of this chapter, AJE was in support of the Muslim Brotherhood in Egypt to the point that was strikingly evident in its daily reporting to audiences across the Arab world. With this particular story of the sentencing of 528 members, this support was even more evident, given also that on the same day three of Al Jazeera's journalists were on trial in an Egyptian court: Peter Greste, Mohamed Fahmy and Baher Mohamed. This direct involvement with Egypt's judicial system means that Al Jazeera is somewhat a direct actor in this story. Whilst it could be argued that such a story lends itself to reporting on the topic with sympathy towards the plight of the detained members of the Muslim Brotherhood and the Al Jazeera journalists, there is still a need to provide both sides of the story as noted in the Johnston case where although the BBC was itself part of the story it remained relatively clear on its role as a mediator too. AJE's reporting on this story is not strongly one-sided, however it fails to provide a voice for the court when narrating the story – something CNNI did as I shall go on to discuss.

The AJE report on the day of the sentencing highlights that the events marks 'the largest number of people to be sentenced to death by a court in Egypt and world history'. This is a big claim, and serves to accentuate the impact of the ruling by using lexical items such as 'world history'. This form of overtextualisation provides a superlative at a global level and across all times. It worth noting that this imposed exaggeration that is stripped from any factual basis, is similar to that of the last AJE report discussed in Chapter 4 on the medics' case, whereby claims were made without any factual backing. Another similarity is that this report on the sentencing of members of the Muslim Brotherhood was also filed by an Arab reporter, an indication that a cultural influence could be at play here.

Notably, CNNI that day described the court ruling in its opening sentence as 'unprecedented in Egypt's history' where 'never has there been such a large capital case'. CNNI confines this case to Egyptian history and not to that of the world, like AJE. BBC, however, is more reserved and uses a spokesperson from Amnesty International to make this point, whereby the speaker states: 'This is the largest number of people sentenced to death in a single trial that Amnesty International has documented.' The BBC's quote is even more specific and factual, and uses an expert on the topic as the invited talent to comment. AJE presents the narrative through an exaggerated claim, in comparison to CNNI and the BBC, and by doing so it emphasises the importance and impact of the event.

AJE was the only channel out of the three that mentioned that day 'the presiding judge was angry and decided to hand down the judgment in two days time', thus insinuating that the court sentence was an impulsive decision by the judge as a result of his personal anger. This is an implication rather than explicitly stated. It is notable that the AJE reporter does not use a source to make this claim and does not provide context as to why the judge was angry, instead he is keen to highlight the recklessness of the court decision. This is very similar to AJE's reporting on the medics' case in the report filed by the Arab reporter, where the Arab journalist is making claims without evidence such as eye witness accounts.

AJE also mentions that 'defence lawyers said they had been given no time to review trial documents'. This is a component of the story that is also mentioned in the BBC report through a quote from one of the sentenced members of the Muslim Brotherhood who were on bail. The member of the Muslim Brotherhood in the BBC report states that the defence team did not have enough time to go over the one thousand page legal documents. It is interesting to note that again AJE includes this point in its reporting without a direct quote or in reference to a certain speaker, whereas the BBC shows the member of the Muslim Brotherhood speaking, although his face is not shown. The lack of evidenced reporting in the AJE report is again noted in this example, in comparison to the BBC.

In regards to the general reporting of the BBC that day, the narrative was somewhat sympathetic to the 528 sentenced. For example, one BBC report described a scene from court as 'the Brotherhood members are getting used to being caged in court'. The lexical item 'caged' along with the accompanying image of the prisoners in a cage, provides a verbal-visual correspondence (Chouliaraki, 2006) of the prisoners, emphasising an inhumane treatment. It is notable that the caption shown at the bottom of the screen during the BBC report reads 'Mass Egypt death sentences "breach international law"'. Although the report itself does not address this in detail, it does state that: 'The trial is taking place but without many of the key defendants, only 62 are in court, most are being tried in absentia'. This is shown against a backdrop of images of the wives and mothers of the defendants. The AJE report does not show these images, which is interesting to note given that as an Arab broadcaster, which is only reporting form the side of the members of the Muslim Brotherhood, it would have been expected that they would have included coverage on the families of the detained. Yet the BBC's reporting on the families in this case is consistent with its style of reporting on the families of the Bulgarian nurses in Chapter 4, whereby the BBC was the only news network to include footage on the Bulgarian nurses' families. Hence, a culture of journalism that is particular to the BBC is becoming apparent here.

Interestingly, CNNI that day was the only channel to provide a right of reply to the court: 'Now the government has defended the trial saying that the sentence was issued by an independent court after careful study of the case.' The AJE and BBC reports did not provide any statement from the court in their coverage. At the same time, CNNI presents a strong argument in support of the Muslim Brotherhood: 'Hundreds of Muslim Brotherhood supporters have died, and thousands have been arrested since last July – that's when a military coup was triggered by massive street demonstrations that ousted Morsi – no one has been held accountable for those deaths'. This logic aims to highlight the injustice of the judicial system in Egypt, a point that was not made, for example, by AJE. In this sense, CNNI provides an account of the event that gives a voice to both sides of the story. It is worth noting that CNNI also highlights that many of the convicted have claimed that 'they were no where near the police station when the attack occurred'. These points all provide additional information not present in the AJE and BBC reports studied, thus providing a more holistic understanding of the event.

Further discrepancies between all three channels can be seen through their ending of the reports. AJE and BBC both state that this court sentence would lead to escalations. For example AJE states: 'The verdict is seen as a sign of the government's widening crackdown on opposition voices, especially members of the outlawed Muslim Brotherhood, raising concerns that this verdict could lead to further instability and violence in the country'. The BBC report also uses a similar ending to that of AJE, where it states that: 'Many see the verdict as a serious escalation of the state's attempt to crush the Brotherhood, and end its opposition on the streets – tomorrow there is another mass trial of the Brotherhood supporters'. It can also be seen that both AJE and BBC are basing these ending statements on anonymous sources, where AJE says that 'The verdict is seen as', but does not state by whom, and the BBC uses 'Many see'. These examples could arguably be regarded as a form of modality whereby the reporter is 'attempting to persuade us of their representation of reality' (Haig, 2001), since this sense of 'instability and violence' as suggested by AJE and 'escalation' as mentioned by the BBC is not reported based on facts, rather it is speculative. This modality becomes further obvious when compared to the CNNI coverage whereby the reporter provides a more realistic consequence to the court sentencing: 'While 529 are sentenced to death, Egypt has a long legal process with many avenues of appeal – it is unlikely that 529 will actually be executed once the sentences are carried out'. This ending statement offers a realistic framing of the consequences of the event, and provides the viewer with more information rather than assumptions.

With these similarities and differences established through the thematic structure that was applied to all three studied global news networks, the main question in relation to this study remains: to what extent can AJE be seen as challenging the content of reports produced by BBC and CNNI? Again in this instance AJE's reporting did not offer a clear 'challenge', rather AJE's reporting is more aligned with the findings of Chapter 4, whereby AJE is reporting from a 'national' angle, given Qatar's political position on the events in Egypt discussed at the start of this chapter. Also, given that Al Jazeera's own reporters were also detained, AJE ultimately held a position that was not in favour of the Egyptian judicial system. It can be argued that AJE presented a similar discourse to that of BBC on this story, yet in comparison to CNNI's coverage which offered a voice to all parties involved, AJE's reporting is evidently one-sided. This example also emphasises a point discussed in Chapter 4 that reveals that CNNI's discursive reporting is different to that of the BBC, therefore debates in the literature which regard these news networks as 'mainstream' or dominant', although descriptive, is not accurate. The research conducted in this book thus far has evidenced that discursively these news networks are very different, and in turn cannot be lumped under the same discursive practice. With this in mind, I will move on to the next case which although is a short analysis, provides an example of AJE's discursive challenge.

Table 5.2   Common thematic items of the Syria UN veto story used by all three channels

Common Thematic Discourse Items
Veto of UN Security Council resolution on Syria
US reaction
Russia's reaction
China's reaction

Syria UN Veto

On 19 July 2012, The UN Security Council in New York failed to adopt a UN resolution on Syria. The Russian Federation and China – which are both key allies to the Syrian government – vetoed a British-sponsored resolution that would have imposed economic sanctions against the Syrian government for failing to carry out a peace plan. This peace plan was negotiated by Kofi Annan who was the special envoy to the UN and the Arab League at the time, and which was ignored by the Syrian President. The resolution that day received 11 votes in favour to two against which were China and the Russian Federation, with two abstentions which were Pakistan and South Africa. The resolution also called for extending the UN Mission's mandate for 45 days which expired on 20 July 2012. It 'would have had the Council act under Chapter VII of the United Nations Charter to demand verifiable compliance – within 10 days of the adoption – with its demands in previous resolutions that Syrian authorities pull back military concentrations from population centres and cease the use of heavy weaponry against them' (Security Council 6810th Meeting, accessed 29 April 2014). This was the third time that Russia and China had vetoed a resolution pertaining to Syria since the Syrian uprisings as part of the Arab Spring. As a result of the Security Council's vote, the 300-member UN mission was left in limbo. From the reports analysed by all three channel the common thematic items arise as follows.

As noted from the above thematic outline, all three channels shared the same components of the narrative. However, interestingly, the way each channel went about reporting on the UN council's meeting was slightly different. The CNNI report was quite straight forward in that it included a clear quote/reaction from each of the countries involved highlighting their position, these countries included Russia, China, the US, the UK, in addition to the UN Secretary General. Note that these sources included in the CNNI coverage were more than those included in Table 5.2. Interestingly, out of the three news networks, CNNI was the most comprehensive in its reported narrative, giving equal time for each position and country. The only additional element in the CNNI report that went beyond the above thematic items, is that the report began by noting that the resolution was vetoed by the UN Russian Ambassador 'along with his Chinese colleague', thus highlighting that Russia and China were a united front.

The BBC report included more insight into what happened during the Council's meeting. The report begins by saying that 'European countries and the US had backed the resolution'. The BBC is the only news network to mention Europe's involvement since, as we have noted from the medics' case, the BBC focuses on the European position on the story due to the UK being part of the European Union (EU). The report goes on to address the US's position by including a quote by the US Ambassador to the UN, Susan Rice, denouncing the outcome of the vote. Interestingly, the BBC report is the only news network out of the three which does not include a British quote or reaction, rather it focuses on the US position, labelling its report in connection to the US: 'US 'disgust' at Russia and China veto on Syria' thus providing a framing for its coverage.

One point which is of particular relevance to this study is in relation to the US delegation walking out during the UN council's meeting as a response to the Syrian UN Ambassador's speech. It is worth noting that CNNI did not mention this incident given that it is the US delegation that walked out. The BBC reported on it as follows: 'During his speech the Syrian Ambassador criticised the US, leading to this response [image of US delegation standing and walking out] The US delegation walking out'. The BBC report did not provide more context to this, whereas AJE does by clarifying that the US delegation had walked out as a result of the Syrian UN Ambassador accusing the US of protecting Israeli interests: 'The Syrian Ambassador accused the US of meddling in Syria to advance Israel's aims prompting the American delegation to walk out'. It is notable that although both the BBC and AJE reported on the US walkout, AJE made reference to the Israeli connection. Examples such as these which resonate with Arab discourse, whereby Arabs generally view the US as protecting Israeli political interests, illustrate how AJE is offering examples of slight challenge to the global mainstream discourse. It is also worth noting, in accordance with the results from Chapter 4, that this report that mentions the Israeli connection with the US was filed by an Arab reporter – whereas another AJE report which was analysed on the Syrian veto that day and which was filed by a Western reporter, did not mention this incident at all.

Although this story is short and contained, it is an example of how AJE is exhibiting a moment of discursive challenge in comparison to BBC and CNNI, whereby AJE is including an Arab discourse within the mediated 'global public sphere' that would usually not be part of global mainstream discourse. In the first story on the sentencing of the Muslim Brotherhood by Egyptian courts, AJE reported on the story with a focus on the plight of the Muslim Brotherhood. This reinstates the findings in Chapter 4 where each channel reports on the event from its national framework. This will be discussed in the following chapter as a factor when considering power relations in the global news space. The difference within AJE itself was also evident in this chapter, whereby the AJE Arab reporter offered an Arab discourse whereas the AJE Western reporter didn't, thus stressing that the culture of journalism a reporter comes from or adopts also plays a role in the news discourse. To add, the BBC and CNNI were providing differing discourses in their reporting; this highlights that dominant mainstream global news networks are not uniform in their discursive output. All these findings will be discussed in the following final chapter, by labelling the factors that affect the discursive output of these news networks as 'forces of discursive media power'.

CHAPTER 6

FORCES OF DISCURSIVE MEDIA POWER

Chapters 4 and 5 sought to look into debates surrounding Al Jazeera English (AJE) and its role both in the Arab and 'global public spheres' before and after the Arab Spring protests. Debates surrounding AJE prior to its launch were concerned with whether AJE would be an 'alternative' news network like Al Jazeera Arabic (AJA) or whether it would operate as a 'watered-down' version of its parent channel. Debates after the Arab Spring witnessed a shift in arguments, whereby as a result of AJA adopting a more political stance on the reporting of the protests through the Qatari political position, discussions circulated around whether it would provide an 'alternative' voice to AJA, which was taking a more mainstream role within the Arab public sphere. Although this book does not empirically address these debates, the changes in discussions positioning AJE in relation to its parent channel before and after the Arab Spring, reflect the fluidity in discursive reporting that news networks can adopt based on factors or 'forces of discursive media power', which this chapter considers in detail. This fluidity within one news organisation and its hybrid tendency for discursive reporting highlights how audiences also play a factor in how news discourses are designed to shape and influence social reality within their sphere of operation.

From the empirical study this research has found, and in turn argues, that the mediated global public space is complex which ultimately places limitations on conceptualising the field through global binaries of North/South, alternative/mainstream, and flow/contra-flow. The empirical analysis conducted in this book suggests that the discursive news coverage of global news networks seems to be contextual on the news story, and in turn depends on the ideological and regional position each news network adopts in relation to that news story. Here, also, the claim made by news networks on offering a global representation such as AJE's self-promotion of providing 'Every Angle / Every Side', and 'All the News / All the Time', is an ideal that does not lend itself to realistic practice as this study has revealed. Accordingly, AJE's hybrid discursive practice of *domination and resistance* seems to better describe the news output of AJE.

With this argument in mind, I will now turn to a more detailed discussion of the research findings specifically in relation to AJE. The research inquiry that has been guiding this study addresses two areas of debate: the first pertaining to the question of whether AJE offers, or may be regarded as embodying, an 'alternative' media platform which offers counter-discourses, thus contesting dominant media power; and the second relates to the nature of the role AJE could be seen to play within a mediated 'global public sphere' as a result. Accordingly, the empirical study of CDA has attempted to provide responses to these two research questions. It is necessary to note that the results of the empirical study do not claim to be representative and offer a comprehensive response, especially through a limited number of news samples. However, what the findings allow me to do is to engage in some depth with the research inquiry through a comparative analysis that offers suggestive responses on how AJE maybe conceptualised within the global news space, especially that the research findings were consistent across all the study samples analysed.

This book has drawn on work that suggests that, within the field of global satellite television news, there are elements that influence the construction of news discourses and in turn discursive media power. Accordingly, I label these elements as 'forces of discursive media power' in the global news arena, and which have been identified in this study as the routines of (1) *news sourcing*, (2) the nature of the *media organisation* itself and its relationship with its audiences, and (3) the influence of the *culture of journalism* on reporting practices. I shall now turn to each of these factors before reflecting on the role of AJE within the 'global public sphere' through the notion of *domination and resistance*.

News Sourcing

It is notable from this study that there were similarities found in the *sourcing of news* across all three channels studied, for example, it was found that, across both sets of stories, the discourses broadcast by AJE, BBC and CNNI were mainly similar, and shared a considerable amount of overlap. Examples of this could be found in the main outline of the stories analysed, in addition to the inclusion of identical footage such as that of Johnston's release or scenes of the European delegation comforting the accused nurses in the medics' case. Yet despite this observation the discourses presented by each news network also differed, indicating that selectivity lies at the core of a news organisation's discursive output. The framing of a news event is thus, in part, an outcome of news organisations presenting a certain discourse through a process of selecting and highlighting the relevant facts, and which also serve to promote the preferred discourse.

AJE's coverage of the medics' story and the UN's veto for Syria provided examples of how AJE's news discourses could often be quite distinct from those of BBC and CNNI, thus providing instances of challenge. Yet through the empirical study, AJE did not always seem to source 'new' facts, since its arguments often lacked supporting evidence. For example, in the medics' story, analysis conducted on AJE found that it did not appear to source additional or different information that supported the 'challenging' discourse it presented. Traditionally, AJA has operated as a two-tier news institution like CNNI as exhibited through its coverage of the Iraq war where it both sourced and delivered its news stories. However, in the report on the medics' case identified in this study as adopting a counter-discourse to the BBC and CNNI, through its accusation of the 'West' of deliberately infecting the Libyan children with the HIV virus, AJE did not seem to source strong factual information to support the position it adopted. We might ask why, then, AJE did not present new information for this report out of the three broadcasters, particularly since it might appear to have privileged access to the story by virtue of language and geography? It should be said here that it was notable that none of the studied channels presented much in the way of evidence on the medics' story, particularly the question of proof presented at the trial, reflecting a wider inadequacy in information sourcing. However, it can be argued that, in order for AJE to offer a substantive challenge to predominant narratives surrounding the issue, there was a need to offer the global viewer background information regarding the Libyan leadership's position, for example, or the Palestinian doctor's role, and probably even Arab responses to the European and American positions. Although the AJE report offered a strong and notable challenge to the discourses broadcast by BBC and CNNI, its claims remained, in the absence of factual evidence, mere assertions.

Here, perhaps, AJE, does share an attribute that has been associated with alternative media, but one which does not position it strongly as a 'contra-flow'. To illustrate, in a study of news sourcing by alternative online websites, Boyd-Barrett found that a majority of online news websites actually relied on 'mainstream US and overseas sources' (Boyd-Barrett, 2007: 180). He notes that in relation to alternative sites, 'the appearance of originality was not so much the product of in-house investigations as the result of loaded linking devices, judicious juxtapositions, and summary evaluations, of stories from diverse sources' (Boyd-Barrett, 2007: 180). The reason for this, according to Boyd-Barrett, is due to alternative sites having minimal newsgathering resources in comparison to mainstream media. As a result, he argues, alternative sites tend to 'reframe' the information they acquire from mainstream media, and do this by taking the original frame which is constructed as a 'final product by mainstream media, and create a new frame where the "final" product becomes a component brick in the construction of a new frame' (Boyd-Barrett,

2007: 180). One concern he raises about this 'reframing' and 'redefining' of the news story originally featured in the mainstream, as Boyd-Barrett argues, is that it could serve to 'consolidate mainstream power' (Boyd-Barrett, 2007: 183). Boyd-Barrett's suggestion that most alternative sites tend to 'reframe' the information acquired from mainstream news agencies or media, rather than undertake different processes of news sourcing is significant to this study's findings on AJE, since original pieces of information presented through mainstream media are merely *reframed* by alternative media, which seems to be the case in the AJE report.

Although this suggests that there is a reliance on similar sourcing which in turn may indicate that there is a similarity in reporting, there still seems to be an adaptation to popular local sentiment as in the case of the AJE report. Accordingly, this discussion suggests that it remains vital to consider the need for alternative news agencies when conceptualising possibilities of increased discursive representation. While there have been attempts by alternative news agencies to provide alternative news, these attempts have confronted issues that impact their operation and survivability due, mainly, to the domination of major international news agencies. Accordingly, alternative news agencies, similar to alternative media, will still have to solicit a more prominent role in global media. There have been attempts by NGOs and intergovernmental institutions to fund national news agencies from the developing world or alternative news agencies (Boyd-Barrett, 2000: 9), such as the example of the Inter Press Service (IPS), the largest global news agency that focuses on news from the developing world (Giffard and Rivenburgh, 2000: 12). IPS, whose motto is 'Journalism and Communication for Global Change', has been operating from Rome, Italy. The IPS website claims that 'IPS raises the voices of the South and civil society [...] IPS tells the story underneath' (Inter Press Service, 23 October 2008), thus clearly situating itself as a challenge to dominant global news agencies. Yet, such examples remain quite limited in the face of traditional global news agencies. In sum, the sourcing of news plays an important role in the formation of media power, and 'what agencies do and how they do it are important for the survival of a "public sphere" of democratic dialogue' (Boyd-Barrett and Rantanen, 2004: 42).

While, like other satellite broadcasters, AJE is likely to remain influenced by the forms of media power exercised by dominant international news agencies, the question of whether it can come to develop as an alternative source of news definition remains one for further consideration and research. The findings of this study, however, suggest it is difficult to argue that it performs such a role already. The concentrated nature of international news agencies, as discussed in Chapter 1, places limitations on discursive variety, which in turn impacts variety in discursive representation. Diversified representation, therefore, is not only required at a media retail level but also at a news sourcing level. Nonetheless, although there is a standardisation of news sourcing, selectivity based on the practice of discursive media power remains central in the construction of the mediated discourse. This was noted in the analysis where the selection of images, quotes, and lexical items all helped shape the sourced material towards a distinctive position on the news story. What this means is that despite a standardisation of news sourcing as a result of the monopoly exercised by international news agencies, international news agendas are adapted and reframed for a regional context by media retailers. This selection of the mediated discourse is therefore shaped by the nature of the *media organisation*, as I shall go on to discuss.

The Media Organisation

While, as discussed above, the influence of international news agencies may tend to produce a standardisation in the production of news, processes of selection based on the *political-economy of the media organisation* in relation to economic and political affiliations, remain an area around which significant discursive differences can emerge. This featured prominently in the construction of discursive media power, in this study, since it is the ideological stance of the news network which constitutes and dictates the nature of the mediated discourse that is being advocated. The analysis uncovered various discursive differences which support this, for example CNNI selected the lexical item 'shadowy' to describe Johnston's kidnappers, whereas AJE used the word 'fringe' to describe the group. This reflects CNNI's negative framing of the Palestinian kidnappers through the terrorism framing, whereas AJE opted to apply a more politically neutral labelling as an Arab broadcaster. In addition, the selection of who to invite to comment on each story also appeared to be influenced by the particular discursive framing presented by the news organisation, such as the use of the quote taken by the Director of Contagious Diseases Centre in Libya for the AJE report on the medics' case, in which he stated that the infection of the children with the AIDS virus was a Western conspiracy. This directly contrasts with the BBC and CNNI reports, which did not include this discourse at all. Indeed, across all three news channels, processes of selection significantly contributed to constructions of social reality in ways that largely correspond to the political cultures of the regions from which each news network originates.

Politically, the media act as an instrument through which global issues and events are both identified and presented in the form of framed, packaged news to the public, and thus exercise power in setting the global political agendas and the parameters of public opinion. Thus, as Schudson argues, '[i]n a political democracy, the media are a vital force in keeping the concerns of the many in the field of vision of the governing few' (Schudson, 1995: 20). This suggests that the media play a vital political role, a point which can be drawn on in support of approaches that suggest news media organisations should be regarded as political institutions that play a key role in the political field because of their ability to construct politics, and also because they follow standard practices in the production of political news (Cook, 2006: 161, Sparrow 2006). In addition, Herbert Schiller also emphasises the extent of economic power media institutions have, noting that '[t]he film and television industries, important since their inception as providers of popular diversion and the dominant ideology, have become economic heavyweights as well, exporting billions of dollars' worth of symbolic product' (Schiller, 2000: 122). These political and economic factors play a role in the determination of discourses each media organisation presents.

Through the thematic structure applied in this study, the second step of *generalisation* – which allowed for the emergence of the thematic narrative – revealed some notable ideological differences in the discourses presented by each news network. Such differences include CNNI's framing of Johnston's kidnappers as terrorists by politically linking them to Al Qaeda. Other differences that emerged can be noted in BBC's political position on the medics' case where it clearly adopted a European position. Such examples are likely, at least in part, to be underpinned by the different political-economic influences that act upon each broadcaster. In order to explore these differences in political economy, it is useful to consider the perspective of Oliver Hahn (2007), who emphasises the importance of considering 'the geography of news' in global journalism and in turn how it affects political mediated discourses. As Hahn argues:

> Despite all the tendencies towards the globalisation of journalism, the geography of news broadcasting remains decisive in the sphere of crisis communication. It therefore seems useful to investigate the politico-cultural foundations for different perspectives and their influence on the selection and presentation of news (Hahn, 2007: 19).

Hahn's work is of particular relevance to this discussion, as he specifically addresses the cultural context of Arab and Western journalism practice, arguing that '[c]ontemporary Arab and Western satellite newscasters do not operate within hermetically sealed spaces but within their respective politico-cultural contexts, and within the communication systems of their respective target audiences. That is one reason why reporting perspectives on the same conflicts can be extremely different' (Hahn, 2007: 20). This argument was supported through the analytic study where, for example, in the Johnston story BBC focused not only on Johnston's release but also on Palestinian/ British diplomatic relations; CNNI strongly adopted the terrorism discourse which, at the time of the story's broadcast, was high on the US agenda; and AJE focused more closely on the Palestinian leadership's effort to secure Johnston's release through a local Arab influence. Similarly, through the medics' case BBC World framed the plight of the Bulgarian nurses as a 'European' concern; CNNI was keen to highlight the AIDS discourse which is the aspect of the story which, arguably, is most newsworthy for US audiences and political culture; whereas AJE focused on the Libyan children as 'local' victims within the Arab world. The inclusion of these discourses was not by chance; rather they are a logical reflection of the political economy of each of these channels on the event at hand. The reports analysed by each channel were dependent on the nature of the news organisation that they stemmed from, which takes into account government influences, national political debates, assured audience interest, and other issues such as the basis of funding. Such influences, including those of Shoemaker and Reese's (1996) hierarchy of influences discussed in Chapter 1, can strongly influence how discursive media power is constructed. The emerging differences in reporting are therefore understood through the institutional agenda of each channel which emerges as a consequence of its particular economic, political and cultural location.

This perspective also leads us to question whether global news networks can be regarded as practising 'global' representation and providing global discourses in the 'comprehensive' manner in which global news networks often claim to do. This may, again, be related to Couldry's (2003) claims regarding the myth of the media being representative of society's centre, which in the discourse of global satellite broadcasting comes to be extended to a rhetorically projected 'global society'. As Couldry argues, the media's constant claims 'to represent society as a whole are almost always ideological in this general sense, and are intrinsic to media institutions' self-image' (Couldry, 2003: 46). Notably, aspects of AJE's marketing, such as the characterisation of the station addressing 'Every Angle / Every Side', and 'All the News / All the Time' serve to both appropriate and reproduce this myth.

Despite this claim to comprehensive coverage it was interesting to note that, through the comparative CDA conducted on the three channels, there were differences in focus and perspective not only in AJE's coverage, but also between BBC and CNNI's coverage. In a study of the 'organisational culture' of the BBC and CNN, Lucy Küng-Shankleman found that there were 'strong similarities between the culture of an organisation and that of its "host" nation' (Küng-Shankleman, 2003: 95). BBC World is shaped through its institutional role as a British public service broadcaster, and CNNI as a global/US commercial broadcaster. Indeed, the BBC coverage on the stories analysed was significantly different to that of CNNI. While it does not follow that references to 'mainstream' or 'dominant' media are meaningless, since it is undoubtedly the case that some centres of media production exercise significantly greater power to shape international perspectives on social reality than others, this serves to emphasise the important point that the 'mainstream' is neither singular nor monolithic.

This point, along with the lack of clear distinction between mainstream and alternative sectors and discourses previously discussed, serve as a basis upon which attempts to categorise broadcasters as either 'mainstream' or 'alternative' might themselves be questioned. This might, by extension, also lead us to question notions of flows/contra-flows where AJE positions itself as a counter-flow to 'Western' or dominant media. This positioning assumes a linear relationship whereby dominant and counter-balancing media operate at opposing ends of the spectrum. Yet what this study suggests is that the global news field is a complex space which cannot be realistically reflected in broad labels such as North/South, flow/contra-flow, and mainstream/alternative.

Culture of Journalism

Another influence which can be regarded as another factor when considering discursive media power, as suggested by this study, is the culture of journalism which was found to play a role in the discursive construction of news within AJE's coverage. Indeed, the analyses noted that differences in AJE's own discursive reporting coincided with the background of the reporters, such that Arab reporters tended to either provide in-depth reporting on Arab politics and discourses which would usually be of interest to Arab audiences as in the

Johnston story, or align their stories more strongly with a popular Arab discourse which was particularly evident in the medics' story and the UN veto story. These examples suggest that the culture of the journalist plays a role in the construction of discursive media power, where the discourse is affected by who is reporting, such that differences in reporting might be related to the different cultural backgrounds of reporters. Here 'culture' not only refers to the ethnic background of the journalist, but also to both the reporter's exposure to the particular political culture of the Arab world and his/her socialisation within the field of Arab journalism, particularly as practiced by AJA. For example, the use of rhetorical tropes used by the Arab reporters identified in the study, whereby journalists use language as a play on words, is used extensively in Arab journalism and forms part of the culture – this is very different from the direct and simple language used in Anglo-Saxon journalism cultures.

Various 'journalism cultures' have been identified in previous studies, where despite there being 'an all-encompassing consensus among journalists toward a common understanding and cultural identity of journalism', different 'professional ideologies' are articulated through journalism cultures (Hanitzsch, 2007: 368). Shoemaker and Reese (1996) suggest that the journalist's opinions and attitudes have an effect on the media messages that he/she produces. Although this influence is at the bottom of their 'hierarchy of influence' on news production, they argue that journalists' 'personal attitudes may translate into selections that undermine the political legitimacy of the covered person or event' (Shoemaker and Reese, 1996: 264). They argue that the inclusion of the journalist's personal attitude relates to the role that the reporter sees him – or herself undertaking, '[w]hether journalists see their roles as interpreting what others do, disseminating information, or serving as an adversary of the powerful, these roles may determine how they define their jobs, the kinds of things they believe should be covered, and the ways in which they cover them' (Shoemaker and Reese, 1996: 264).

At a global level, the varying cultures of journalism become distinctly visible. Mark Deuze (2002) has investigated various news cultures through a consideration of different national contexts, and accordingly defines national news culture 'as an intervening variable between people (cf. journalists, sources, or publics) and a given "objective" situation (cf. media events, organizations, infrastructures, and systems) through which citizens inform or are informed, can be seen as partly carried by the broadly defined and operationalized profile of media professionals within a given national context' (Deuze, 2002: 134). Hence, to borrow his conceptualisation of news culture, a *national* news culture consists of various factors which include the journalists who work within that culture, the various kinds of storytelling, and their relation with sources and the public (Deuze, 2002: 134). All these characteristics, Deuze suggests, are articulated and connected through a national context, which in turn affects the making of the news. Indeed, Deuze argues that while it could be assumed that journalists working within the global news arena might have a 'shared set of characteristics' thus forming what he labels as an 'international news culture', when differences arise they 'can be attributed to national context, as an indicator of particularity' (Deuze, 2002: 135). This was partially evident in the analytic study, which did suggest that AJE often exhibited a differing form of journalism practice to that of the BBC and CNNI and, in addition, suggested that different forms of journalism practice may also co-exist within AJE itself. For example, in considering the AJE Arab reporter for the medics' story, the use of strong emotional appeals and personal rhetoric which favoured a particular Arab discourse could suggest, for example, that the journalist's personal Arab disposition played a role in shaping the presented mediated discourse. This, in turn, resulted in a discourse that was distinct from the non-Arab AJE reporter on the same story. Again, it might be argued that this finding is related to a tension in AJE's managerial strategy, which attempts to strike a 'balance' of novelty and legitimacy within the international arena by adopting journalistic styles, and apparently also similar news values, to those of their global competitors, whilst also presenting Arab positions and concerns. In the course of the analysis, the possibility that two distinct 'cultures of journalism' co-exist within AJE's textual output was raised, on the grounds that AJE has recruited both Arab reporters who have previously worked for its parent channel, while also hiring reporters from the international field of news broadcasting. In this respect, it is worth considering the potential role that the distinctive histories, career trajectories and cultures of journalism together play in contributing to AJE's hybrid news output.

These findings bear upon a broader consideration of claims regarding whether AJE can be understood as performing a 'counter-public sphere' role within the global news space. Due to the differences in discursive reporting within AJE itself, it is difficult to argue on the basis of the empirical study that AJE functions as a counter-public sphere in a strong sense since AJE appears, in many respects, 'undecidable' within a framework of media power. Indeed it may be the case that AJE serves to partially, if somewhat weakly, contest dominant sites of media power. The findings suggest a greater complexity. This complexity can be understood as a product of AJE's hybrid nature and that the consideration of this case study, in turn, highlights the complexity of power relations within the field of satellite television news.

The difficulty in any straightforward categorisation of AJE, that is, can be linked to the numerous, and somewhat contradictory, influences on discursive media power discussed above that serve to inform AJE's identity and practice. Indeed, the analysis suggests that each media channel reports in a manner that is largely consistent with political and economic factors that are likely to be influencing its coverage, such as the diplomatic stance of the government in the country in which the news organisation is based, the local and regional concerns related to actors implicated in each story, and the dominant agendas of public debate within different polities and news markets. In this regard, it was evident that all three news networks analysed were producing news discourses which promoted a certain position consistent with their institutional, political and cultural location, and that this in turn served to influence their practice of discursive media power. This leads to the second problem raised by this research, which is the extent to which AJE could be considered as playing a role in the realisation of a mediated 'global public sphere'.

AJE's Hybrid Practice

The three positions discussed in the opening chapter of this book on the existence of a 'global public sphere' within the field of international news circulate around the issue of representation at a global level. Representation is essential in the formation of a democratic and inclusive mediated global space yet, as Sparks (1998) argues, there are significant obstacles that stand in the way of the realisation of a comprehensive global news representation. Such limitations include state restrictions and (a frequent lack of) secure funding. However, it could be argued that, in the case of AJE, both these limitations have been overcome. With the launch of AJE in English, the barrier of language which Sparks raises as a major issue is overcome, at least to the extent that its audience is not restricted to a local audience defined by language.

With this in mind, it is necessary to understand the nature of AJE's role as a recent entrant to the field of global news through debates on the mediated 'global public sphere' that were discussed in Chapter 1. The debates on the question of a 'global public sphere' outlined various positions on whether developments in satellite broadcasting, including both the emergence of global broadcasters and an increase in channels from various parts of world should be understood as a 'democratisation' of the mediated 'global public sphere' or, alternatively, an expansion of neo-imperialism. Discussion of these debates noted that, particularly among arguments that posit an increasingly democratised field of global news is being facilitated by satellite television, Al Jazeera has been frequently referenced as an exemplary case to provide support for this position. We may recall, however, Sakr's argument (2007a) that 'contra-flow in its full sense would seem to imply not just reversed or alternative media flows, but a flow that is also counter-hegemonic' (Sakr, 2007a: 105). This suggests that the issue is not just related to the capacity of emergent players to access global audiences, but the degree to which they present discursive output that represents a significant and consistent 'contra-flow' that contests the terms of dominant media representation.

The results of this study, however, suggest that AJE's inclusion in the 'global public sphere' may not provide a consistent and significant challenge to discourses broadcast by the BBC and CNNI. During the course of the analytic study there were numerous instances where it was difficult to distinguish AJE reports from those of BBC and CNNI. In a sense this is not quite surprising since, as mentioned at the start of this book, AJE self-avowedly set out to emulate the two Western news networks (El-Nawawy and Iskandar, 2002). However, the way in which AJE promoted itself as a counter-balance to Western media may have created an expectation that the majority of its reports would provide differing perspectives to those offered elsewhere. Instead, the reports produced by AJE examined in this study are, on the whole, quite similar in content to that of CNNI and BBC.

We have also seen, however, that while AJE largely reproduces similar accounts of events to those presented by BBC and CNNI, it does on occasion offer viewers different and distinctive perspectives on these. This could be regarded as an attempt by AJE to represent views from the Arab world by showcasing the political, social and economic discourses found in the Arab world in relation to the news event. Although reports analysed in this study were not a direct contestation to the discourses found in reports by the BBC and CNNI, they arguably did provide an insight into Arab political mediated discourses. In this respect, although AJE did not firmly challenge Western discourses, it may have provided an alternative means of interpreting events that might enable alternative understandings of events to circulate globally.

Yet, the expectation that AJE might strongly realise a 'free market of ideas' is, thus, one that this book has not supported: in fact, its analysis has suggested that AJE more than often confirms existing discourses within the dominant sphere. The expectation that AJE fulfils this ideal is one that, in part, has resulted from AJA's role in expanding available news discourses in the Arab world. However, AJA was broadcasting to a specific audience and in an environment which was quite limited in terms of liberal, free reporting and debate. This study has found that AJE provides a partial challenge; and in the examples where this challenge was detected it did not provide factual evidence based reporting, instead it advocated a position through emotional appeals and often conspiracy theories. Nonetheless, AJE's performance in the global news space can be articulated through the notion of a hybrid practice, whereby its news reporting often serves to advocate positions and perspectives which reflect its institutional ideology as an Arab broadcaster.

AJE's performance, therefore, can be regarded as an example of Arjun Appadurai's description of global culture where 'sameness' and 'difference' blend to produce the 'triumphantly universal and the resiliently particular' (Appadurai, 1990: 307–308). This quality, Appadurai suggests, emerges as a consequence of disjunctive flows of people, technologies, money, ideologies and media within what he describes as the 'global cultural economy' that is marked by a 'tension between cultural homogenization and cultural heterogenization' (Appadurai, 1990: 295). This tension has been exemplified through the empirical study conducted in this book, where AJE's coverage has been shown to contain elements shared in common with its global competitors as well as elements that are distinctive and even, on occasion, contest the terms of their discourses. The findings of this study have, in short, identified examples of news reporting that reflected 'sameness' and 'difference' simultaneously. Furthermore, and as a result of a simultaneous flow of homogenisation *and* heterogenisation, Appadurai suggests that this new global cultural economy 'cannot any longer be understood in terms of existing center-periphery models (even those that might account for multiple centers and peripheries)' (Appadurai, 1990: 296).

It is notable that strong arguments that claim a 'global public sphere' is increasingly being realised and arguments that suggest that the political and economic realities of global media provision make such a claim implausible, both rest on a strongly idealist perspective, wherein the public sphere is presented as either an ideal that is being realised, or critical standard of democratic communication against which the inadequacies of contemporary communicative politics can be diagnosed. By contrast, Fraser (1993) suggests that the concept can be drawn upon to consider the 'actually existing' public sphere, as one that is constituted partially through exclusion, but which remains a continual terrain of struggle around which 'counter-publics' form, and struggle to develop communicative resources, and gain access to and transform broader public arenas of communication. Yet it can be argued that, while it offers a more sophisticated model for considering the politics of public communication, this model also offers an inadequate basis for considering communicative power in the case of AJE. Situating AJE within Fraser's model is problematic since, through its performance of both 'sameness' and 'difference', it raises questions about the adequacy of Fraser's model, which appears to rest on a binary opposition between a dominant 'mainstream' public sphere and counterpublics which are positioned as discrete, external spaces of communicative activism. There is, ultimately, limited scope within this model to allow us to consider cases such as AJE which produce a hybrid practice. As we have seen, AJE can be regarded as mainstream in the sense that it exists as a commercial broadcaster that seeks to attract a global audience, and which has adopted a very traditional, mythic discourse of 'comprehensiveness', and produces news that appears to share many, if not most, of its discursive features in common with its global 'mainstream' competitors. At the same time, we have also seen occasions when AJE's coverage did offer significantly different interpretations of news events in ways that, even if relatively weakly evidenced, may yet provide significant variations to predominant perspectives.

This hybrid practice suggests that the inclusion of additional voices from previously unrepresented parts of the world into the mediated 'global public sphere', while significant, does not guarantee that 'contestation' will be a subsequent result. Moreover, it is also notable that the reporting of the BBC and CNNI, as dominant global broadcasters, is not itself uniform in its representation. What this serves to highlight is, again, a complexity that is belied by mainstream/alternative, flow/contra-flow, and North/South binaries. Rather there is a need to position media organisations through their social, political and economic associations, thus reflecting the various journalistic influences that affect the discursive constructions of international news networks. These various influences produce a complex news activity.

As noted at the beginning of this book, a particular importance of this research lies in the fact that it reports on what is a unique and emergent global broadcaster that explicitly claims to represent a challenging perspective, presenting different discourses, and reporting back in the face of dominant news flows. It is understandable, following long standing concerns over dominant flows of news from 'North' to 'South', that the emergence of a global channel that lays claim to such a role has led to some excitement, particularly as it appears to represent the first well-funded news channel with such an agenda. This has led researchers to suggest that AJE might represent a strikingly important development that provides a radically different perspective on news. This book has sought to address this in two ways: firstly, through an attempt to develop a more substantive engagement with what it might mean to suggest that Al Jazeera is 'alternative'; and secondly, through a qualitative engagement with the discursive form of AJE's news coverage, contrasting this with its major global rivals.

In this final chapter, which attempts to make sense of what appeared as contradictory findings that emerged from the analysis, it was argued that different forces of discursive media power serve to influence and shape the nature of the discursive output of international news networks. What this study highlights is that it is not possible, within the global news space, for one news channel to offer a comprehensive account of a news event, rather audience habits would need to adopt multi-news network viewing routines. One former Al Jazeera reporter noted post-Arab Spring that:

Today, Arab media is divided. Media outlets have become like parties; politics dominates the business and on both sides of the landscape and people can't really depend on one channel to get their full news digest. It is as if the audience have to do journalists' homework by cross-checking sources and watching two sides of a conflict to get one piece of news (Hashem, 2012).

The findings of this book do not necessarily suggest that the nature of news organisations adopting a certain discourse when covering a news event is problematic, as long as the 'global public sphere' is inclusive of global media representation.

This book addresses television coverage in isolation as a focus of this study, yet AJE has a significant web platform and this also merits further study, as does the central place of the internet for discussions on the 'global public sphere'. This study suggests that the emergence of new media markets may, in some ways, enable new voices to gain access to the 'global public sphere' in order to practice discursive media power, an area that warrants further research. Further research which uncovers other discursive practices in television satellite news could shed light on the findings of this study. For example, comparative studies on other emergent global broadcasters from what were previously 'silent' parts of the world could also assist in addressing the research inquiry, as it would be worth understanding whether there are similar discursive patterns to those found within AJE from other new entrants to the mediated 'global public sphere'.

BIBLIOGRAPHY

The Age (2006) 'Al-Jazeera Reveals Australia Plans', 17 November 2006 [Online] http://www.theage.com.au/news/tv–radio/aljazeera-reveals-australia-plans/2006/11/17/1163266754914.html, accessed 13 March 2007.

Allam, Hannah (2006) 'New Al-Jazeera International Channel Sparks Conflict', in *News Center – Breaking News and Views for the Progressive Community*, CommonDreams.org, 2 March 2006 [Online] http://www.commondreams.org/headlines06/0302-08.htm, accessed 22 August 2007.

Appadurai, Arjun (1990) 'Disjuncture and Difference in the Global Cultural Economy', in *Theory, Culture & Society*, Vol. 7, No. 2, pp. 295–310.

Atton, Chris (2002) *Alternative Media*, Sage Publications: London.

———— and James F. Hamilton (2008) *Alternative Journalism*, Sage Publications: London.

Azran, Tal (2004) 'Resisting Peripheral Exports: Al Jazeera's War Images on US Television', in *Media International Australia: Incorporating Culture and Policy*, No. 113, November 2004, pp. 75–86.

Baker, Aryn (2011) 'Bahrain's Voiceless: How al-Jazeera's Coverage of the Arab Spring is Uneven', in *Time*, 24 May [Online] http://world.time.com/2011/05/24/bahrains-voiceless-how-al-jazeeras-coverage-of-the-arab-spring-is-uneven/, accessed 23 April 2014.

Barkho, Leon (2007) 'Unpacking the Discursive and Social Links in BBC, CNN and Al-Jazeera's Middle East Reporting', in *Journal of Arab and Muslim Media Research*, Vol. 1, No. 1, pp. 11–29.

BBC News, 'Al-Jazeera English TV Date Set', 1 November 2006 [Online] http://news.bbc.co.uk/go/pr/fr/-/2/hi/middle_east/6105952.stm, accessed 22 June 2007.

Benson, Rodney (2004) 'Bringing the Sociology of Media Back In', in *Political Communication*, No. 21, pp. 275–292.

Blommaert, Jan and Chris Bulcaen (2000) 'Critical Discourse Analysis', in *Annual Review of Anthropology*, No. 29, pp. 447–466.

Boaden, Helen (2004) 'This is the BBC News' [online] http://news.bbc.co.uk/newswatch/ukfs/hi/newsid_3970000/3975900/3975913.stm, accessed 12 May 2011.

Bob, Clifford (2008) 'Conservative Forces, Communications and Global Civil Society: Towards Conflictive Democracy', in *Media Spaces: Innovation and Activism*, in *Global Civil Society 2007/8: Communicative Power and Democracy*, Clifford Bob, Jonathan Haynes, Victor Pickard, Thomas Keenan and Nick Couldry, pp. 198–203.

Boyd-Barrett, Oliver and Daya Kishan Thussu (1992) *Contra-Flow in Global News: International and Regional News Exchange Mechanisms*, John Libbey & Co: London.

——— (1997) 'Global News Wholesalers as Agents of Globalization', in *Media in a Global Context: A Reader*, Annabelle Sreberny-Mohammadi, Dwayne Winseck, Jim McKenna and Oliver Boyd-Barrett (Eds), Arnold: London, pp. 131–144.

——— (2000) 'National and International News Agencies: Issues of Crisis and Realignment', in *Gazette*, Vol. 62, No. 1, Sage Publications: London, pp. 5–18.

——— (2002) 'Theory in Media Research', in *The Media Book*, Chris Newbold, Oliver Boyd-Barrett and Hilde Van Den Bulck (Eds), Arnold: London, pp. 1–54.

——— and Terhi Rantanen (2004) 'News Agencies as News Sources: A Re-Evaluation', in *International News in the Twenty-First Century*, Chris Paterson and Annabelle Sreberny (Eds), John Libbey Press: London, pp. 31–46.

——— (2007) 'Alternative Reframing of Mainstream Media Frames', in *Media on the Move: Global Flow and Contra-Flow*, Daya Kishan Thussu (Ed.), Routledge: London, pp. 178–194.

Calhoun, Craig (1992) 'Introduction: Habermas and the Public Sphere', in *Habermas and the Public Sphere*, Craig Calhoun (Ed.), MIT Press: Massachusetts, pp. 1–48.

Carlsson, Ulla (2003) 'The Rise and Fall of NWICO – and Then?: From a Vision of International Regulation to a Reality of Multilevel Governance', EURICOM Colloquium in Venice 5–7 May 2003, Information Society: Visions and Governance.

Carvalho, Anabela (2008) 'Media (ted) Discourse and Society', in *Journalism Studies*, Vol. 9, No. 2, pp. 161–177.

Castells, Manuel (2009) *Communication Power*, Oxford University Press: Oxford.

CCR (2006) 'No Carrier: Al Jazeera International to be Available only Online in the U.S.', in *Arab Media*, posted 24 July 2006 [Online] http://arab-media.blogspot.com/2006/07/no-carrier-al-jazeera-international-to.html, accessed 22 August 2007.

Chalaby, Jean K. (2005) 'Towards an Understanding of Media Transnationalism' in *Transnational Media Worldwide: Towards a New Media Order*, Jean K. Chalaby (Ed.), I.B.Tauris: London, pp. 1–13.

Chouliaraki, Lilie (2006) 'Towards An Analytics of Mediation', in *Critical Discourse Studies*, Vol. 3, No. 2, pp. 153–178.

Clausen, Lisbeth (2003) *Global News Production*, Copenhagen Business School Press: Copenhagen.

Cook, Timothy E. (2006) 'The News Media as a Political Institution: Looking Backward and Looking Forward', in *Political Communication*, Vol. 23, No. 2, pp. 159–171.

Corner, John (2000) 'Influence: The Contested Core of Media Research', in *Mass Media and Society* (3rd Edn), James Curran and Michael Gurevitch (Eds), Arnold: London, pp. 376–397.

Cotter, Colleen (2001) 'Discourse and Media', in *The Handbook of Discourse Analysis*, Deborah Schiffrin, Deborah Tannen and Heidi E. Hamilton (Eds), Blackwell Publishing: Oxford, pp. 416–436.

Cottle, Simon (2003) 'Media Organisation and Production: Mapping the Field', in *Media Organisation and Production*, Simon Cottle (Ed.), Sage Publications: London, pp. 3–24.

————— (2006) *Mediatized Conflict: Developments in Media and Conflict Studies*, Open University Press: Berkshire.

————— (2009) 'Global Crisis in the News: Staging New Wars, Disasters, and Climate Change', in *International Journal of Communication*, Vol. 3, pp. 494–516.

————— and Mugdha Rai (2008) 'Global 24/7 News Providers: Emissaries of Global Dominance or Global Public Sphere?', in *Global Media and Communication*, Vol. 4, No. 2, pp. 157–181.

Couldry, Nick (2000) *The Place of Media Power: Pilgrims and Witnesses of the Media Age*, Routledge: London.

————— (2002) 'Mediation and Alternative Media, or Relocating the Centre of Media and Communication Studies', in *Media International Australia incorporating Culture and Policy: Citizens' Media*, Christina Spurgeon and Ellie Rennie (Eds), No. 103, May, pp. 24–31.

————— (2003) 'Beyond the Hall of Mirrors? Some Theoretical Reflections on the Global Contestation of Media Power', in *Contesting Media Power: Alternative Media in a Networked World*, Nick Couldry and James Curran (Eds), Rowman & Littlefield: Oxford, pp. 39–54.

————— and James Curran (2003a) 'The Paradox of Media Power', in *Contesting Media Power: Alternative Media in a Networked World*, Nick Couldry and James Curran (Eds), Rowman & Littlefield: Oxford, pp. 3–15.

————— (2010) *Why Voice Matters: Culture and Politics after Neoliberalism*, Sage: London.

Cunningham, Stuart, Elizabeth Jacka and John Sinclair (1998) 'Global and Regional Dynamics of International Television Flows', *Electronic Empires: Global Media and Local Resistance*, in Daya K. Thussu (Ed.) London: Arnold, pp. 177–192.

————— (2004) 'Popular Media as Public "Sphericules" for Diasporic Communities', in *The Television Studies Reader*, Robert C. Allan and Annette Hill (Eds), Routledge: London and New York, pp. 151–162.

Curran, James (1991) 'Rethinking the Media as a Public Sphere', in *Communication and Citizenship: Journalism and the Public Sphere*, Peter Dahlgren and Colin Sparks (Eds), Routledge: London, pp. 27–57.

Cushion, Stephen (2010) 'Three Phases of 24-Hour News Television', in *The Rise of 24-Hour News Television: Global Perspectives*, Stephen Cushion and Justin Lewis (Eds), Peter Lang: New York, pp. 15–29.

Dana, Rebecca (2005) 'Al That Jaz!', in the *New York Observer*, 5 December 2005.

Demers, David (2002) *Global Media: Menace or Messiah?*, Hampton Press: NJ.

Deuze, Mark (2002) 'National News Cultures: A Comparison of Dutch, German, British, Australian and U.S. Journalists', in *Journalism and Mass Communication Quarterly*, Spring, Vol. 79, No. 1, pp. 134–149.

Donnelly, Sally B. (2005) 'Al Jazeera Hires an Ex-Marine', in *Time*, 27 September 2005 [Online] http://www.time.com/time/nation/article/0,8599,1110008,00. html, accessed 22 August 2007.

Downing, John D. H. (1995) 'Alternative Media and the Boston Tea Party', in *Questioning the Media: A Critical Introduction* (2nd Edn), John D. H. Downing, Ali Mohammadi and Annabelle Sreberny (Eds), pp. 238–252, Sage Publications: California.

———— (2001) *Radical Media: Rebellious Communication and Social Movements*, Sage Publications: London.

Edwards, Haley Sweetland (2011) 'Former Al Jazeera Head on Quitting, the Arab Spring, and Qatar's Role', in *The Atlantic*, 30 September [Online] http://www.theatlantic.com/international/archive/2011/09/former-al-jazeera-head-on-quitting-the-arab-spring-and-qatars-role/245932/, accessed 23 April 2014.

El Amrani, Issandr (2006) 'In US, a Cold Welcome for Al Jazeera International', in *Transnational Broadcasting Studies*, No. 16, Fall [Online] http://tbsjournal.com/ElAmrani.html, accessed 22 August 2007.

ElGhul-Bebawi, Saba (2009) 'The Relationship between Mainstream and Alternative Media: A Blurring of the Edges?', in J. Gordon (Ed.) *Notions of Community: A Collection of Community Debates and Dilemmas*, Bern: Peter Lang, pp. 17–32.

El-Nawawy, Mohammed and Adel Iskandar (2002) *Al-Jazeera: How the Free Arab News Network Scooped the World and Changed the Middle East*, Westview Press: Cambridge, MA.

———— and Adel Iskandar (2002a) 'The Minotaur of "Contextual Objectivity": War Coverage and the Pursuit of Accuracy with Appeal', in *Transnational Broadcasting Studies*, No. 9, Fall/Winter [Online] http://www.tbsjournal.com/Archives/Fall02/Iskandar.html, accessed 3 September 2010.

———— and Shawn Powers (2010) 'Al-Jazeera English: A Conciliatory Medium in a Conflict-driven Environment?', in *Global Media and Communication*, Vol. 6, No. 1, April, pp. 61–84.

Eley, Geoff (1992) 'Nations, Publics, and Political Cultures: Placing Habermas in the Nineteenth Century', in *Habermas and the Public Sphere*, Craig Calhoun (Ed.), MIT Press: Massachusetts, pp. 289–339.

Entman, Robert M. (2009) 'Framing Media Power', in *Doing News Framing Analysis: Empirical and Theoretical Perspectives*, Paul D'Angelo and Jim Kuypers (Eds), Routledge: Hoboken, pp. 331–355.

Fairclough, Norman (2001) *Language and Power*, Longman: Harlow.

Figenschou, Tine Ustad (2010) 'A Voice for the Voiceless?: A Quantitative Content Analysis of Al-Jazeera English's Flagship News', in *Global Media and Communication*, Vol. 6, No. 1, April, pp. 61–84.

———— (2013) *Al Jazeera and the Global Media Landscape: The South is Talking Back*, Taylor and Francis: Hoboken.

Flew, Terry (2007) *Understanding Global Media*, Palgrave Macmillan: New York.

Flood, Chris, Stephen Hutchings, Galina Miazhevich and Henri Nickels (2011) 'Between Impartiality and Ideology', in *Journalism Studies*, Vol. 12, No. 2, pp. 221–238.

Flyvbjerg, Bent (2006) 'Five Misunderstandings About Case-Study Research', in *Qualitative Inquiry*, Vol. 12, No. 2, pp. 219–245.

Fowler, Roger, Bob Hodge, Gunther Kress and Tony Trew (1979) *Language and Control*, Routledge: London.

Fraser, Nancy (1993) 'Rethinking the Public Sphere: A Contribution to the Critique of Actually Existing Democracy', in *The Phantom Public Sphere*, Bruce Robbins (Ed.), University of Minnesota Press: Minneapolis, pp. 1–32.

———— (2007) 'Transnationalizing the Public Sphere: On the Legitimacy and Efficacy of Public Opinion in a Post Westphalian World', in *Identities,*

Affiliations, and Audiences, Seyla Benhabib, Ian Shapiro and Danilo Petranovic (Eds), Cambridge University Press: Cambridge, pp. 45–66.

——— (2007a) 'Transnational Public Sphere: Transnationalizing the Public Sphere: On the Legitimacy and Efficacy of Public Opinion in a Post-Westphalian World', in *Theory, Culture & Society'*, Vol. 24, No. 7, Sage Publications: London, pp. 7–30.

——— (2009) *Scales of Justice: Reimagining Political Space in a Globalizing World*, Columbia University Press: New York.

Fuchs, Christian (2010) 'Alternative Media as Critical Media', in *European Journal of Social Theory*, Vol. 13, No. 2, pp. 173–192.

Fürsich, Elfriede (2009) 'In Defense of Textual Analysis', in *Journalism Studies*, Vol. 10, No. 2, pp. 238–252.

Giffard, Anthony C. and Nancy K. Rivenburgh (2000) 'News Agencies, National Images, and Global Media Events', in *Journalism and Mass Communication Quarterly*, Vol. 77, No. 1, pp. 8–21.

Gitlin, Todd (1998) 'Public Sphere or Public Sphericules', in *Media Ritual and Identity*, Tamar Liebes and James Curran (Eds), Routledge: London, pp. 168–174.

Golding, Peter and Phil Harris (1997) 'Introduction', in *Beyond Cultural Imperialism: Globalization, Communication & the New International Order*, Peter Golding and Phil Harris (Eds), Sage Publications: London, pp. 1–9.

Guaaybess, Tourya (2002) 'A New Order of Information in the Arab Broadcasting System', in *Transnational Broadcasting Studies*, No. 9, Fall/Winter [Online] http://www.tbsjournal.com/Archives/Fall02/Guaaybess.html, accessed 9 September 2007.

Habermas, Jürgen (1990) *Moral Consciousness and Communicative Action*, Cambridge: MIT Press.

——— (1992) *The Structural Transformation of the Public Sphere: An Inquiry into a Category of the Bourgeois Society*, trans. by Thomas Burger with the assistance of Frederick Lawrence, Polity Press: Cambridge.

——— (1992a) 'Further Reflections on the Public Sphere', in *Habermas and the Public Sphere*, Craig Calhoun (Ed.), MIT Press: Massachusetts, pp. 421–461.

Hackett, Robert A. and Yuezhi Zhao (1998) *Sustaining Democracy? Journalism and the Politics of Objectivity*, Garamond: Toronto.

Hahn Oliver (2007) 'Cultures of TV News Journalism and Prospects for a Transcultural Public Sphere', in *Arab Media and Political Renewal: Community, Legitimacy and Public Life*, Naomi Sakr (Ed.), I.B.Tauris: New York, pp. 13–27.

Haig, Edward (2001) 'A Study of the Application of Critical Discourse Analysis to Ecolinguistics and the Teaching of Eco-literacy', in *Studies in Language and Culture*, Vol. 22, No. 2, pp. 205–226.

Hamilton, James (2000) 'Alternative Media: Conceptual Difficulties, Critical Possibilities', in *Journal of Communication Inquiry*, Vol. 24, No. 4, October, pp. 357–378.

Hanitzsch, Thomas (2007) 'Deconstructing Journalism Culture: Toward a Universal Theory', in *Communication Theory*, No. 17, pp. 367–385.

Harcup, Tony (2003) 'The Unspoken – Said: The Journalism of Alternative Media', in *Journalism: Theory, Practice and Criticism*, Vol. 4, No. 3, pp. 356–376.

Hashem, Ali (2012) 'The Arab Spring has Shaken Arab TV's Credibility', in *Guardian*, 3 April [Online] http://www.theguardian.com/commentisfree/2012/apr/03/arab-spring-arab-tv-credibility, accessed 23 April 2014.

Hiro, Dilip (2006) 'Why the World Needs Al-Jazeera English', in *Guardian*, 20 November 2006 [Online] http://commentisfree.guardian.co.uk/dilip_hiro/2006/11/post_662.html, accessed 22 August 2007.

Hussain, Ghaffar (2013) 'The Collapse of Al-Jazeera's Credibility', in *The Commentator* [Online] http://www.thecommentator.com/article/2741/the_collapse_of_al_jazeera_s_credibility, accessed 23 April 2014.

Inter Press Service, 'Get To Know Us' [Online] http://www.ips.org/institutional, accessed 23 October 2008.

Iskandar, Adel (2006) 'Is Al Jazeera Alternative? Mainstreaming Alterity and Assimilating Discourses of Dissent', in *Transnational Broadcasting Studies*, No. 15, Spring [Online] http://www.tbsjournal.com/Iskandar.html, accessed 6 October 2006.

Ito, Mizuko (2008) 'Introduction', in *Networked Publics*, Kazys Varnelis (Ed.), MIT Press: Cambridge, pp. 1–14.

Jenkin, Mathew (2006) 'Do Not Adjust Your Sets: An Alternative View of the World', in *Independent*, Reported by Guy Adams, 15 November 2006 [Online] http://news.independent.co.uk/media/article1984447.ece, accessed 22 August 2007.

Jewkes, Yvonne (2004) *Media and Crime*, Sage Publications: London.

Johnson, Pauline (2006) *Habermas: Rescuing the Public Sphere*, Routledge: London and New York.

Kaldor, Mary (2003) *Global Civil Society: An Answer To War*, Polity: Malden, MA.
——— and Helmut K. Anheir and Marlies Glasius (2005) 'Introduction', in *Global Civil Society 2004/2005*, Helmut K. Anheier, Mary Kaldor and Marlies Glasius (Eds), Sage Publications: London, pp. 1–25.

Kessler, Oren (2012) 'The Two Faces of Al Jazeera', in *The Middle East Quarterly*, Winter, pp. 47–56 [Online] http://www.meforum.org/3147/al-jazeera, 23 April 2014.

Khamis, Sahar (2007) 'The Role of New Arab Satellite Channels in Fostering Intercultural Dialogue: Can Al Jazeera English Bridge the Gap?', in *New Media and the New Middle East*, Philip Seib (Ed.), Palgrave Macmillan: New York, pp. 39–51.

Khanfar, Wadah (2012) 'Al Jazeera and the Arab Spring', Talk at Chatham House, London, 19 January.

Kitzinger, Jenny (2007) *Media Studies: Key Issues and Debates*, Eoin Devereux (Ed.), Sage Publications: London, pp. 134–161.

Kühn, Alexander, Christoph Reuter and Gregor Peter Schmitz (2013) 'After the Arab Spring: Al Jazeera Losing Battle for Independence', in *Spiegel Online* [Online] http://www.spiegel.de/international/world/al-jazeera-criticized-for-lack-of-independence-after-arab-spring-a-883343.html, accessed 23 April 2014.

Küng-Shankleman, Lucy (2003) 'Organisational Culture Inside the BBC and CNN', in *Media Organization and Production*, Simon Cottle (Ed.), Sage Publications: London, pp. 77–96.

Landes, Joan B. (1988) *Women and the Public Sphere in the Age of the French Revolution*, Cornell University Press, New York.

Locke, Terry (2004) *Critical Discourse Analysis*, Bloomsbury: London.

Lorenzo-Dus, Nuria (2009) *Television Discourse: Analysing Language in the Media*, Palgrave Macmillan: UK.

Lull, James (2007) *Culture-on-Demand: Communication in a Crisis World*, Blackwell Publishing: MA.

Lynch, Marc (2006) *Voices of the New Arab Public: Iraq, Al-Jazeera, and Middle East Politics Today*, Columbia University Press: New York.

Masie, Elliot (2006) 'CNN Newsroom in the Midst of Katrina', in *Public Library Quarterly*, Vol. 24, No. 2, pp. 73–76.

McChesney, Robert W. (1999) *Rich Media, Poor Democracy: Communication Politics in Dubious Times*, University of Illinois Press: Urbana and Chicago.

———— (2003) 'Corporate Media, Global Capitalism', in *Media Organization and Production*, Simon Cottle (Ed.), Sage Publications: London, pp. 27–39.

McGuigan, Jim (1996) *Culture and the Public Sphere*, Routledge: London and New York.

McNair, Brian (2006) *Cultural Chaos: Journalism, News and Power in a Globalised World*, Routledge: London and New York.

Miles, Hugh (2005) *Al-Jazeera: The Inside Story of the Arab News Channel that is Challenging the West*, Grove Press: New York.

Montgomery, Martin (2007) *The Discourse of Broadcast News: A Linguistic Approach*, Routledge: New York.

Murdock, Graham and Peter Golding (2005) 'Culture, Communications and Political Economy', in *Mass Media and Society* (4th Edn), James Curran and Michael Gurevitch (Eds), Hodder Arnold: London, pp. 60–83.

Mytton, Graham (2008) 'The BBC and its Cultural, Social and Political Framework', in *Historical Journal of Film, Radio and Television*, Vol. 28, No. 4, pp. 569–581.

Negt, Oskar and Kluge, Alexander (1993) *Public Sphere and Experience: Toward an Analysis of the Bourgeois and Proletarian Public Sphere*, University of Minnesota Press: Minneapolis.

Newman, Edward (2007) *A Crisis of Global Institutions?: Multilateralism and International Security*, Routledge: Hoboken.

Nisbet, Erik C. and Myers, Teresa A. (2010) 'Challenging the State: Transnational TV and Political Identity in the Middle East', in *Political Communication*, Vol. 27, pp. 347–366.

Painter, James (2008) *Counter-hegemonic News: A Case Study of Al-Jazeera English and Telesur*, Reuters Institute for the study of Journalism, University of Oxford.

Parks, Lisa (2003) 'Our World, Satellite Televisuality, and the Fantasy of Global Presence', in *Planet TV: A Global Television Reader*, Lisa Parks and Shanti Kumar (Eds), New York University Press: New York, pp. 74–93.

Paterson, Christopher (1997) 'Global Television News Services', in *Media in a Global Context: A Reader*, Annabelle Sreberny-Mohammadi, Dwayne Winseck, Jim McKenna and Oliver Boyd-Barrett (Eds), Arnold: London, pp. 145–160.

People's Daily Online, 'Al-Jazeera Plans to Set Alternative Agenda', in *People's Daily Online*, source: China Daily, 6 July 2005 [Online] http://english.people.com.cn/200507/06/eng20050706_194443.html, accessed 3 August 2007.

Peters, Bernhard (2010) 'National and Transnational Public Spheres', in *The Idea of the Public Sphere: A Reader*, Jostein Gripsrud, Hallvard Moe, Anders Molander and Graham Murdock (Eds), Lexington Books: Plymouth, pp. 237–246.

Philo, Greg (2007) 'Can Discourse Analysis Successfully Explain the Content of Media and Journalistic Practice?', in *Journalism Studies*, Vol. 8, No. 2, pp. 175–196.

Pickard, Victor W. (2007) 'Neoliberal Visions and Revisions in Global Communications Policy from NWICO to WSIS', in *Journal of Communication Inquiry*, Vol. 31, No. 2, pp. 118–139.

Pintak, Lawrence (2005) 'Interview with Nigel Parsons, Managing Director of Al Jazeera International', in *Transnational Broadcasting Studies*, No. 15, 23 October 2005 [Online] http://www.tbsjournal.com/Parsons.html, accessed 13 September 2007.

———— (2006) 'Al-Jazeera International, Not Quite Ready for Takeoff', in *Columbia Journalism Review*, 27 April 2006 [Online] http://www.cjr.org/behind_the_news/aljazeera_international_not_qu.php, accessed 22 August 2007.

Powers, Shawn and Mohammed El-Nawawy (2009) 'Al-Jazeera English and Global News Networks: Clash of Civilizations or Cross-cultural Dialogue', in *Media, War & Conflict*, Vol. 2, No. 3, pp. 263–284.

———— and Eytan Gilboa (2007) 'The Public Diplomacy of Al Jazeera', in *New Media and the New Middle East*, Philip Seib (Ed.), Palgrave Macmillan: New York, pp. 53–80.

Rai, Mugdha and Simon Cottle (2007) 'Global Mediations: On the Changing Ecology of Satellite Television News', in *Global Media and Communication*, Vol. 3, No. 51, Sage Publications: London, pp. 51–78.

Rantanen, Terhi (2005) *The Media and Globalization*, Sage Publications: London.

Reese, Stephen D. (2001) 'Understanding the Global Journalist', in *Journalism Studies*, Vol. 2, No. 2, pp. 173–187.

Richardson, John E. (2007) *Analysing Newspapers: An Approach from Critical Discourse Analysis*, Palgrave Macmillan: New York.

Rinnawi, Khalil (2006) *Instant Nationalism: McArabism, Al-Jazeera and Transnational Media in the Arab World*, University Press of America: Oxford.

Robinson, Piers (2011) 'The CNN Effect Reconsidered: Mapping a Research Agenda for the Future', in *Media, War & Conflict*, April, No. 4, pp. 3–11.

Rodriguez, Clemencia (2001) *Fissures in the Mediascape: An International Study of Citizens' Media*, Hampton Press Communication Series: Cresskill, NJ.

Rugh, William, A. (2004) *Arab Mass Media: Newspapers, Radio, and Television in Arab Politics*, Praeger: Westport.

Sakr, Naomi (2005) 'Maverick or Model? Al-Jazeera's Impact on Arab Satellite Television', in *Transnational Television Worldwide: Towards a New Media Order*, Jean K. Chalaby (Ed.), I.B.Tauris: London, pp. 66–95.

———— (2007) *Arab Television Today*, I.B.Tauris: London.

———— (2007a) 'Challenger or Lackey?: The Politics of News on Al-Jazeera', in *Media on the Move: Global Flow and Contra-Flow*, Daya Kishan Thussu (Ed.), Routledge: London, pp. 104–118.

Schiller, Herbert I. (2000) 'Digitised Capitalism: What has Changed?', in *Media Power, Professionals and Policies*, Howard Tumber (Ed.), Routledge: London, pp. 116–126.

Schudson, Michael (1995) *The Power of News*, Harvard University Press: Cambridge.

Security Council, 6810th Meeting (AM), 'Security Council Fails to Adopt Draft Resolution on Syria that would have Threatened Sanctions, Due to Negative

Votes of China, Russian Federation', Department of Public Information, News and Media Division, New York [Online] http://www.un.org/News/Press/docs/2012/sc10714.doc.htm, accessed 29 April 2014.

Seib, Philip (2009) 'Public Diplomacy and Journalism: Parallels, Ethical Issues, and Practical Concerns', in *American Behavioral Scientist*, Vol. 52, No. 5, pp. 772–86.

Shoemaker, Pamela J. and Stephen D. Reese (1996) *Mediating the Message: Theories of Influences on Mass Media Content* (2nd Edn), Longman: New York.

Smith, Philippa and Allan Bell (2007) 'Unravelling the Web of Discourse Analysis', in *Media Studies: Key Issues and Debates*, Eoin Devereux (Ed.), Sage Publications: London, pp. 78–100.

Sonwalker, Prasun (2004) 'News Imperialism: Contra View from the South', in *International News in the 21st Century*, Chris Paterson and Annabelle Sreberny (Eds), John Libbey Publishing: Eastleigh, Hants, pp. 111–126.

Souaiaia, Ahmed E. (2011) 'Qatar, Al Jazeera, and the Arab Spring', in *MRZine*, 17 November [Online] http://mrzine.monthlyreview.org/2011/souaiaia171111.html, accessed 23 April 2014.

Sparks, Colin (1998) 'Global Media: A Global Public Sphere?', in *Electronic Empires: Global Media and Local Resistance*, Daya Kishan Thussu (Ed.), Arnold: London, pp. 108–124.

——— (2005) 'Media and the Global Public Sphere: An Evaluative Approach', in *Global Activism, Global Media*, Wilma de Jong, Martin Shaw and Neil Stammers (Eds), Pluto Press: London, pp. 34–49.

Sparrow, Bartholomew H. (2006) 'A Research Agenda for an Institutional Media', in *Political Communication*, Vol. 23, No. 2, pp. 145–157.

Sreberny-Mohammadi, Annabelle (1997) 'The Many Cultural Faces of Imperialism', in *Beyond Cultural Imperialism: Globalization, Communication & the New International Order*, Peter Golding and Phil Harris (Eds), Sage Publications: London, pp. 49–68.

Straubhaar, Joseph D. and Duarte, Luiz G. (2005) 'Adapting US Transnational Television Channels to a Complex World: From Cultural Imperialism to Localization to Hybridization', in *Transnational Television Worldwide: Towards a New Media Order*, Jean K. Chalaby (Ed.), I.B.Tauris: London, pp. 216–253.

Sullivan, Sarah (2002) 'An Interview with Eason Jordan, CNN Chief News Executive', in *Transnatoinal Broadcasting Studies*, No. 8, [Online] http://www.tbsjournal.com/Archives/Spring02/jordan.html.

Thompson, John B. (1995) *The Media and Modernity: A Social Theory of the Media*, Polity: Oxford.

Thussu, Daya Kishan (2000) *International Communication: Continuity and Change*, Arnold: London.

——— (2004) 'Media Plenty and the Poverty of News', in *International News in the Twenty-First Century*, Chris Paterson and Annabelle Sreberny (Eds), John Libbey Press: London, pp. 47–62.

——— (2005) 'The Transnationalization of Television: The Indian Experience', in *Transnational Television Worldwide: Towards a New Media Order*, Jean K. Chalaby (Ed.), I.B.Tauris: London, pp. 156–172.

——— (2007) 'Mapping Global Media Flow and Contra-Flow', in *Media on the Move: Global Flow and Contra-Flow*, Daya K. Thussu (Ed.), Routledge: London and New York, pp. 10–29.

Ulrichsen, Kristian Coates (2012) *Small States with a Big Role: Qatar and the United Arab Emirates in the Wake of the Arab Spring*, Durham, UK: Al-Sabah Number 3, October 2012.

UNESCO Report by the International Commission for the Study of Communication Problems (1980) 'Communication and Society Today and Tomorrow, Many Voices One World: Towards a New More Just and More Efficient World Information and Communication Order', Kogan Page: London/Unipub, New York/UNESCO, Paris.

Van den Bulck, Hilde (2002) 'Tools for Researching the Media', in *The Media Book*, Chris Newbold (Ed.), Arnold: London, pp. 55–100.

van Dijk, Teun A. (1985) *New Approaches to the Analysis of Mass Media: Discourse and Communication*, Walter de Gruyter: Berlin and New York.

———— (1988) *News as Discourse*, Lawrence Erlbaum Associates Publishers: New Jersey.

———— (1998) *Ideology: A Multidisciplinary Approach*, Sage Publications: London.

———— (2003) 'Critical Discourse Analysis', in *The Handbook of Discourse Analysis*, Deborah Schiffrin, Deborah Tannen and Heidi E. Hamilton (Eds), Blackwell Publishing: MA, pp. 352–371.

Volkmer, Ingrid (1999) *News in the Global Sphere: A Study of CNN and its Impact on Global Communication*, University of Luton Press: Luton.

———— (2002) 'Journalism and Political Crises in the Global Network Society', in *Journalism After September 11*, Barbie Zelizer and Stuart Allan (Eds), Routledge: London, pp. 235–246.

———— (2007) 'Governing the "Spatial Reach"? Spheres of Influence and Challenges to Global Media Policy', in *International Journal of Communication*, Vol. 1, pp. 56–73.

Wodak, Ruth and Michael Meyer (2009) 'Critical Discourse Analysis: History, Agenda, Theory and Methodology', in *Methods of Critical Discourse Analysis* (2nd Edn), Ruth Wodak and Michael Meyer (Eds), Sage Publications: London, pp. 1–33.

Yin, Robert K. (2003) *Applications of Case Study Research* (2nd Edn), Sage Publications: California.

Youssef, Ahmed Magdy (2013) 'On Al-Jazeera's Lopsided coverage of Egypt', in *Open Democracy*, 2 August {Online} http://www.opendemocracy.net/print/74491, accessed 23 April 2014.

Zayani, Mohamed (2008) 'Arab Media, Corporate Communications, and Public Relations: The Case of Al Jazeera', in *Asian Journal of Communication*, Vol. 18, No. 3, pp. 207–222.

———— and Sofiane Sahraoui (2007) *The Culture of Al Jazeera: Inside an Arab Media Giant*, McFarland & Co.: Jefferson, N.C.

INDEX

Printed in Great Britain
by Amazon

32455449R00116